T0290343

Introduction to Stars and Planets

An activities-based exploration

 AMERICAN ASTRONOMICAL SOCIETY  IOP | ebooks™

AAS Editor in Chief

Ethan Vishniac, Johns Hopkins University, Maryland, USA

About the program:

AAS-IOP Astronomy ebooks is the official book program of the American Astronomical Society (AAS), and aims to share in depth the most fascinating areas of astronomy, astrophysics, solar physics and planetary science. The program includes publications in the following topics:

GALAXIES AND COSMOLOGY

INTERSTELLAR MATTER AND THE LOCAL UNIVERSE

STARS AND STELLAR PHYSICS

EDUCATION, OUTREACH, AND HERITAGE

HIGH-ENERGY PHENOMENA AND FUNDAMENTAL PHYSICS

THE SUN AND THE HELIOSPHERE

THE SOLAR SYSTEM, EXOPLANETS, AND ASTROBIOLOGY

LABORATORY ASTROPHYSICS, INSTRUMENTATION, SOFTWARE, AND DATA

Books in the program range in level from short introductory texts on fast-moving areas, graduate and upper-level undergraduate textbooks, research monographs and practical handbooks.

For a complete list of published and forthcoming titles, please visit iopscience.org/books/aas.

About the American Astronomical Society

The American Astronomical Society (aas.org), established 1899, is the major organization of professional astronomers in North America. The membership (~7,000) also includes physicists, mathematicians, geologists, engineers and others whose research interests lie within the broad spectrum of subjects now comprising the contemporary astronomical sciences. The mission of the Society is to enhance and share humanity's scientific understanding of the universe.

Editorial Advisory Board

Steve Kawaler
Iowa State University, USA

Ethan Vishniac
Johns Hopkins University, USA

Dieter Hartmann
Clemson University, USA

Piet Martens
Georgia State University, USA

Dawn Gelino
NASA Exoplanet Science Institute, Caltech, USA

Joan Najita
National Optical Astronomy Observatory, USA

Bradley M. Peterson
The Ohio State University / Space Telescope Science Institute, USA

Scott Kenyon
Smithsonian Astrophysical Observatory, USA

Daniel Savin
Columbia University, USA

Stacy Palen
Weber State University, USA

Jason Barnes
University of Idaho, USA

James Cordes
Cornell University, USA

Introduction to Stars and Planets

An activities-based exploration

Alan Hirshfeld

Department of Physics, University of Massachusetts Dartmouth, North Dartmouth, MA 02747, USA

IOP Publishing, Bristol, UK

© IOP Publishing Ltd 2020

All rights reserved. No part of this publication may be reproduced, stored in a retrieval system or transmitted in any form or by any means, electronic, mechanical, photocopying, recording or otherwise, without the prior permission of the publisher, or as expressly permitted by law or under terms agreed with the appropriate rights organization. Multiple copying is permitted in accordance with the terms of licences issued by the Copyright Licensing Agency, the Copyright Clearance Centre and other reproduction rights organizations.

Permission to make use of IOP Publishing content other than as set out above may be sought at permissions@ioppublishing.org.

Alan Hirshfeld has asserted his right to be identified as the author of this work in accordance with sections 77 and 78 of the Copyright, Designs and Patents Act 1988.

ISBN   978-0-7503-3691-8 (ebook)
ISBN   978-0-7503-3689-5 (print)
ISBN   978-0-7503-3692-5 (myPrint)
ISBN   978-0-7503-3690-1 (mobi)

DOI   10.1088/2514-3433/abc249

Version: 20201201

AAS–IOP Astronomy
ISSN 2514-3433 (online)
ISSN 2515-141X (print)

British Library Cataloguing-in-Publication Data: A catalogue record for this book is available from the British Library.

Published by IOP Publishing, wholly owned by The Institute of Physics, London

IOP Publishing, Temple Circus, Temple Way, Bristol, BS1 6HG, UK

US Office: IOP Publishing, Inc., 190 North Independence Mall West, Suite 601, Philadelphia, PA 19106, USA

*Dedicated to the many thousands of UMass Dartmouth students I have taught
and who have taught me over the years.*

Contents

Preface

My career in astronomy began in a suburban driveway, eight miles outside of New York City. My "observatory" was a small reflector telescope I had purchased with saved-up coins in a large glass jar: $84.50 poured onto the counter of an optical instrument store. On weekends, I'd spend hours under the night sky, inspecting the blurry disk of a planet or the iridescent specks of a binary star. Like so many before me, the teenage astronomy bug proved durable and spurred me through years of training to become a professional in the field. That education swept my perspective on the universe from an earthbound driveway into the depths of space. Those gleaming disks and dots I had observed through my telescope morphed into minutely charted planet-scapes and gigantic blazing suns, and the once-inscrutable workings of the celestial bodies resolved into the universal laws of physics. This book embodies that cognitive leap by presenting the methods by which astronomers investigate the Sun, the stars, and the planets, which now include thousands of exoplanets circling stars beyond our solar system.

The activities in this book were developed over the past five years for my introductory astronomy course, *Stars, Planets, and the Search for Extraterrestrial Life*, here at UMass Dartmouth. This class has proved popular among both science- and non-science majors, as well as students in our University Honors Program. Altogether, these solar-, stellar-, and planetary-themed exercises tackle a core question: how do astronomers know what they know about the universe? Taken in sequence, they illustrate the epic advancement of stellar and planetary astronomy over the past century, up to the present day.

The book is divided into three parts. The first part retraces the pathways by which astronomers measured the basic physical attributes of the Sun—distance, size, mass, temperature, luminosity, chemical composition—before delving into the historical enigma of solar energy production. The second part moves beyond the solar system to the determine the physical characteristics of stars. Here we learn how to "weigh" a star and how to construct and interpret the Hertzsprung–Russell (HR) diagram, which visually portrays the complex evolutionary changes undergone by a star as it ages. Of special interest are the various end stages of stellar evolution: supernova explosions, white dwarfs, and black holes. The second part closes with the realization that the space distribution of globular star clusters reveals the solar system's place within the Galaxy.

The book's third part segues into planetary astronomy, a field that now encompasses the galactic realm, with its multitude of worlds circling other stars. A calendar exercise reveals the disparity between human or societal time scales and the grindingly slow pace of Earth's geological processes and development. From astronomical observations and elementary physics theory, the physical properties of planets are deduced, which are then applied to the concept of habitable zones around stars. The third part concludes with a pair of exercises that illustrate how astronomers discover exoplanets.

The activities are designed to be accessible to college liberal arts majors, yet sufficiently engaging to those in STEM fields. The necessary mathematical tools—basics of computation, geometry, trigonometry, and graphing—are introduced only as needed for each activity. Wherever possible, formulas have been simplified to avoid tedious calculation, yet preserve the inherent relationship among the physical variables. My long experience in the classroom has convinced me that applying quantitative methods to a series of well-defined tasks—and achieving the correct answers!—eases math anxiety, while highlighting the practical benefits of quantitative reasoning.

This book can be used either as a stand-alone core of an activities-oriented astronomy course or as a supplement to an existing textbook. The activities require no specialized equipment or individual materials beyond a pencil, millimeter ruler, and calculator; as such, they are suitable for classrooms of any size, from a seminar room to a lecture hall, as well as for web-based distance-learning courses.

Alan Hirshfeld, Newton, Massachusetts, October 2020

Acknowledgements

This book owes much to my own teachers over the decades, starting with high-school enrichment classes in astronomy at the Newark Museum in New Jersey through my undergraduate and graduate training, where I learned what it takes to be a professional in the field, as well as my obligation to foster succeeding generations of students in their efforts to learn about astronomy. I am also grateful to my many writing mentors who inspired me toward excellence in the books I have written, including this one.

Thanks, too, to the referees who reviewed the manuscript and provided helpful feedback, and to Leigh Jenkins, Sarah Armstrong, and the staff at IOP Publishing for their work bringing the book to completion. My colleagues in the Physics Department at UMass Dartmouth have been a constant source of ideas and insights about teaching, much of which is manifest in the pages that follow. Finally, I thank my wife Sasha, my steadfast companion and biggest fan for the past 46 years.

Author Biography

Alan Hirshfeld

Alan Hirshfeld, Professor of Physics at the University of Massachusetts Dartmouth, is Chair of the American Astronomical Society's Historical Astronomy Division and a longtime Associate of the Harvard College Observatory. He is the author of *Parallax: The Race to Measure the Cosmos*; *The Electric Life of Michael Faraday*; *Eureka Man: The Life and Legacy of Archimedes*; *Astronomy Activity & Laboratory Manual*; and *Starlight Detectives: How Astronomers, Inventors, and Eccentrics Discovered the Modern Universe*. He is a regular book reviewer for the *Wall Street Journal* and writes and lectures frequently on science history and discovery. Visit the author's website at www.alanhirshfeld.com.

Part I

The Sun

Introduction to Stars and Planets
An activities-based exploration
Alan Hirshfeld

Activity 1

The Sun's Distance I: The Method of Aristarchus

Preview

The ancient Greek astronomer Aristarchus used an observation of the Moon to deduce the distance to the Sun. Although he greatly underestimated the solar distance, his methodology was valid and represents one of the earliest efforts to apply geometry to cosmic measurement.

1.1 Aristarchus's Distance to the Sun

Sometime around 250 BCE, the Greek astronomer Aristarchus conceived a way he might deduce the Sun's distance from a particular observation of the Moon. He envisioned how the Earth, Moon, and Sun are arranged in space at the instant when the Moon appears precisely half-illuminated—its first-quarter or last-quarter phase—as pictured in Figure 1.1. At that time, Aristarchus reasoned, the Sun rays strike the Moon perpendicular to our viewing direction from Earth. That is, the Earth, Moon and Sun together form a right triangle, one of whose angles measures 90°, as in Figure 1.2. In this figure, the little square indicates the "right," or 90°, angle; the side labeled r represents the distance from the Earth to the Moon; and the side labeled d—the triangle's hypotenuse—is the distance between the Earth and the Sun. A similar right-triangle arrangement would occur if you held an illuminated ball at arm's length along a direction perpendicular to the direction of the incoming light.

Basic trigonometry provides the means to estimate the Sun's distance. In a right triangle, the cosine of an angle (abbreviated cos) is defined as the length of the side adjacent to the angle divided by the length of the hypotenuse. For the Earth–Moon–Sun triangle in Figure 1.2, this means that $\cos(E) = r/d$, where E is the angle between the Moon and Sun as viewed from the Earth. To envision angle E, imagine standing

doi:10.1088/2514-3433/abc249ch1   1-1   © IOP Publishing Ltd 2020

Figure 1.1. The Moon in its first-quarter phase. (Credit: NASA/GSFC/USRA.)

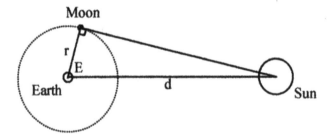

Figure 1.2. Right-triangle configuration of the Earth, Moon, and Sun during the Moon's first-quarter phase.

outside and extending your right arm toward the Sun and your left arm toward the Moon; the angle formed by your outstretched arms is E.

A bit of algebra makes the cosine equation look like this:

$$d = r/\cos(E). \tag{1.1}$$

Having already determined the Moon's distance r, all Aristarchus needed to do to find the Sun's distance was to measure the angle E during the Moon's quarter phase. Easier said than done! If Figure 1.1 were drawn to actual scale, with the Sun very far away compared to the Moon—that is, with d many times larger than r— angle E would be very close to, although not quite, a right angle. (In fact, trigonometry had not been invented in Aristarchus's time; he used analogous geometrical methods to accomplish the procedure described here.)

1. According to an ancient source, Aristarchus estimated—guessed?—that, during the quarter-moon phase, the angle E was 87°. Use this estimate in Equation (1.1) to determine how many times farther away the Sun is than the Moon. Your answer should take the form $d = n \times r$, where n is a number. For now, leave r unspecified; it's merely a symbol that represents the Moon's distance. For example, if angle E were 30°, then cos $(E) = 0.866$; plugging this into Equation (1.1), the Sun's distance d equals $r/0.866$, or $1.15 \times r$.

2. The solar distance d derived from Aristarchus's method is exquisitely sensitive to the value of angle E. In other words, even a slight change in angle E produces a substantial change in the solar distance d. To illustrate, recompute the solar distance if Aristarchus had assumed an angle E merely one degree larger, that is, 88° instead of 87°. Again express your answer in the form $d = n \times r$.

3. Modern measurement reveals that Aristarchus was way off in his estimation of the angle E (so far off, in fact, that we suspect he just plucked a value "out of the hat"). The true value of E is 89.85°. Recompute the solar distance now, again in the form $d = n \times r$.

1.2 The Sun's Diameter

Once the Sun's distance is known, it's easy to find the Sun's true diameter, in units we'll call "Moon-diameters," that is, the number of Moons that would fit across the face of the Sun. During a total solar eclipse, the Moon almost precisely covers the Sun; in other words, the Moon and the Sun *appear* the same angular diameter in the sky. However, the Sun is much farther away than the Moon; therefore, to *appear* the same diameter as the Moon, the Sun must, in fact, be a much larger body than the Moon. For example, if the Sun were three times farther than the Moon, yet appeared the same size as the Moon in the sky, we conclude that the Sun must actually be three Moon-diameters wide. Or in general, if the Sun is n times farther than the Moon, yet they appear the same size, the Sun must be n Moon-diameters wide.

4. (a) Using your value for n from part 1, write down an expression for the Sun's diameter in units of Moon-diameters, according to Aristarchus. (b) Aristarchus went on to find the Sun's diameter in units we will call "Earth-diameters." To do this, he estimated, from lunar eclipse observations, that the Moon is 1/3 as wide as the Earth. Considering this information and your answer to part 4(a), write down an expression for the Sun's diameter in units of Earth-diameters, according to Aristarchus.

1.3 The Sun's Distance Revisited

Curiously, the extant evidence suggests that Aristarchus might not have carried out the next logical step: finding the Sun's distance in units of Earth-diameters. With such information, he could have formed a true scale model of the Sun–Earth system, in the same way that a globe shows a scaled-down version of our planet. The Sun's distance is obtained by answering the following question: given the Sun's true width,

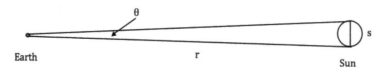

Figure 1.3. Located at distance *r*, the Sun's diameter *s* spans an angle *θ* in the sky.

expressed in Earth-diameters in part 4(b), how many Earth-diameters away must it be to appear as small as it does in the sky—a mere ½ degree across?

To answer this question, we apply a geometrical relation we will call the sector formula:

$$\mathbf{r} = 57.3 \times \mathbf{s}/\boldsymbol{\theta}, \tag{1.2}$$

as illustrated in Figure 1.3. Here *r* is the Sun's distance, *s* is the Sun's true diameter (expressed in Earth-diameters), and the Greek letter theta *θ* is the Sun's apparent diameter in the sky, ½ degree. For example, if an object is 10 Earth-diameters wide (*s*) and spans an angle of 4° in the sky (*θ*), its distance (*r*) is equivalent to about 143 Earth-diameters.

5. Using the sector formula Equation (1.2), with your answer to part 4(b) for the estimated diameter of the Sun, compute the Sun's distance in units of Earth-diameters, according to Aristarchus.

Modern measurements reveal that Aristarchus erred significantly in his estimates of the Sun's diameter and distance: the Sun is actually more than 100 Earth-diameters across and 11,000 Earth-diameters distant! There's nothing wrong with Aristarchus's method; it was just impossible to reliably carry out in his day. Nevertheless, Aristarchus's erroneous solar distance became the *de facto* standard into the Middle Ages. For the first time, someone had calculated the size and distance of a celestial body using observational data gathered from Earth.

Worksheet, Activity 1: The Sun's Distance I: The Method of Aristarchus

Name _____

1.

2.

3.

4. (a)

   (b)

5.

Introduction to Stars and Planets
An activities-based exploration
Alan Hirshfeld

Activity 2

The Sun's Distance II: Transits and Radar-ranging of Venus

Preview

The distance between the Earth and the Sun is determined by observations of the planet Venus: first, by tracking the passage of Venus across the Sun's disk during a rare event called a transit; and, second, by detecting the faint echo of a radar pulse reflected off Venus's surface.

2.1 Transits of Venus

A relatively accurate way of measuring the Sun's distance from the Earth—the astronomical unit, abbreviated au—presents itself whenever either of the inner planets, Mercury or Venus, passes between Earth and the Sun. During these *transits*, the planet is seen in silhouette as a tiny, black disk that creeps across the Sun's luminous face, as depicted in Figure 2.1. Because of the slight tilt of Venus's orbit relative to that of the Earth, its transits occur in pairs, each pair separated by about a century. (Mercury transits thirteen times every century, but it is more difficult to discern its tinier form against the Sun.)

After developing his mathematical laws of planetary motion in the early 1600s, Johannes Kepler derived dates in the future when Mercury and Venus would pass directly between the Earth and the Sun. Mercury's predicted transit of 1631 was observed by French astronomer Pierre Gassendi; Venus's 1639 transit was seen by both Jeremiah Horrocks and William Crabtree in England. (The most recent transits of Venus occurred on 2004 June 8, and 2012 June 5; the next Venus transit-pair won't happen until the years 2117 and 2125.) The English astronomer Edmond Halley realized during the late 17th century that the upcoming transits of Venus in 1761 and 1769 offered the opportunity to determine the astronomical unit by meticulous observation of the planet's silhouette against the Sun's glowing disk.

doi:10.1088/2514-3433/abc249ch2
© IOP Publishing Ltd 2020

Figure 2.1. Transit of Venus (upper left) in 2012, from NASA's Solar Dynamics Observatory spacecraft. (Credit: NASA/SDO, HMI. CC BY 2.0.)

Deducing the Sun's distance by observing a transit of Venus is a three-step process: first find the relative sizes of the Earth's and Venus's orbits; next measure the distance between the Earth and Venus during the transit; and finally combine the preceding data to compute the Sun's distance.

2.2 The Relative Sizes of Earth's and Venus's Orbits

In 1543, Nicolaus Copernicus published his heliocentric cosmic model, which situated the Sun at the center of the universe, with the Earth and the other planets orbiting around it. By measuring the geometry of certain planetary configurations, Copernicus was able to derive the relative spacing of the planets. (Just over a half-century later, Johannes Kepler confirmed and refined Copernicus's estimates.) The procedure for the inner planets, Mercury and Venus, is very simple. Since these planets have orbits smaller than the Earth's, they never stray very far from the Sun in the sky. Venus is never more than 47° away from the Sun in the sky, an angle designated its *greatest elongation*. The geometry of greatest elongation is shown in Figure 2.2, where the Greek letter ϕ ("phi") is the greatest elongation angle, R is the distance of Venus from the Sun, and 1 au is the distance between the Earth and the Sun.

Copernicus realized that when Venus is at greatest elongation ϕ, it forms a right triangle with the Earth and the Sun, where the right angle, indicated in Figure 2.2 by the little square, is located at Venus. This configuration is tailor-made for trigonometry's sine function, defined as the *length of the side opposite an angle* divided by the *length of the hypotenuse*. The definition yields the formula:

$$R = \sin \phi. \qquad (2.1)$$

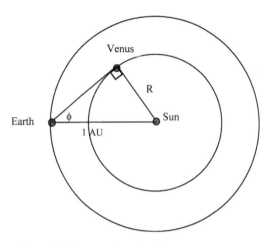

Figure 2.2. Venus at its point of greatest elongation.

1. By substituting the observed greatest elongation of Venus (47°) into Equation 2.1, compute the distance **R** of Venus from the Sun, as Copernicus did. Your answer will be in au.
2. (a) As it moves around its orbit, what is the closest that Venus comes to Earth, in au? (b) Sketch the specific configuration of the Earth, Venus, and the Sun when this closest approach occurs.

2.3 The Absolute Distance from the Earth to Venus

During a transit, observers at different locations on the Earth will see Venus's silhouette at slightly different positions against the Sun's face. This is the familiar parallax effect: the apparent shift in an object's position when viewed from different vantage points. (We experience a parallax effect in observing our surroundings with two eyes.) In Figure 2.3, two Earth-based observers, separated by a distance *s*, are simultaneously viewing the transit of Venus. Observer 1 sees Venus's silhouette at position V1 on the Sun's disk, while Observer 2 sees it slightly offset, at position V2. The angle of this parallax shift is represented by the Greek letter θ ("theta").

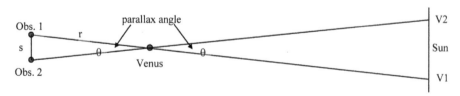

Figure 2.3. Two Earth-based observers viewing a transit of Venus. V1 and V2 represent the respective positions of Venus on the Sun's disk, as seen by Observers 1 and 2.

If the two observers on the Earth measure the parallax angle θ, then the distance to Venus r can be computed from Equation 2.2:

$$r = 206,265 \times s/\theta. \tag{2.2}$$

Here r and s are both expressed in kilometers (km) and the angle θ is expressed in a tiny angular unit called an arcsecond, which is $^1/_{3600}$ of a degree. (The shorthand label for an arcsecond is a "double-prime" symbol: ".)

Figure 2.4 is a picture that shows a superposition of two photographs of the 2004 Venus transit obtained by observers in Essen, Germany, and Uis, a village in Namibia, Africa. These locations are separated by a distance s of **7590 km**. As you can see, the silhouette of Venus appears twice, due to the parallax effect. We can use

Figure 2.4. Merged images of the 2004 June 8, transit of Venus, photographed from Essen, Germany, and Uis, Namibia. (Credit: Backhaus, U. 'Forschen *und* forschendes Lernen beim Venustransit 2004.' http://www. didaktik.physik.uni-due.de/~backhaus/Venusproject/results/CUVenus.pdf.)

this picture to compute the parallax angle θ. However, first we need to determine the picture's scale, that is, how many millimeters on the picture represents an arcsecond of angle.

3. (a) Using a millimeter ruler, measure the diameter of the Sun, to the nearest *tenth* of a millimeter (mm); that is, your measurement should have one decimal place. (b) The angular diameter of the Sun on the day of the transit was observed to be **1890″**. Use your measured solar diameter, along with this observed angular diameter, to compute the picture's scale in units of ″/mm.
4. (a) Using the millimeter ruler, carefully measure the separation between the centers of the two images of Venus, to the nearest *tenth* of a millimeter. (b) Use your picture scale from part 3(b) to compute the parallax angle θ, in arcseconds.
5. Substitute the numbers for the parallax angle θ and the observer-separation s into Equation 2.2 to compute the distance r, in kilometers, between the Earth and Venus.
6. (a) The computed distance r in part 5 corresponds to a specific fraction of an au, which you determined in part 2(a). Use this information to compute the au itself, in kilometers. Explain your reasoning. (b) Compute the percentage difference between your estimate of the au and the au's actual value, which is approximately 149,000,000 km, as follows: [(actual value − estimated value)/ (actual value)] × 100.

2.4 Radar-ranging of Venus

7. The transit method to estimate the au was superseded in 1961 when scientists from the Jet Propulsion Laboratory, in Pasadena, California, measured Venus's distance by reflecting a radar beam off its surface. (A radar pulse travels through space at the speed of light, about 300,000 kilometers per second, or km s^{-1}.) The scientists found that, at Venus's closest approach to the Earth, the radar pulse returned 276.2 seconds (s) after it was sent it out.
   (a) Given this information, describe the steps you would follow to determine the absolute distance to Venus.
   (b) Compute the distance to Venus, in kilometers, using your method.
   (c) From your answer to the previous part, compute the au, in kilometers. Explain how you did it.
   (d) Compute the percentage difference between this radar-based estimate of the au and the au's actual value, as you did in part 6(b).
8. Why do you think scientists didn't try to bounce a radar beam off the Sun itself to determine the au directly?

Worksheet, Activity 2: The Sun's Distance II: Transits and Radar-ranging of Venus

Name _____

1. $R =$ _____ au

2. (a) _____ au, (b)

3. (a) _____ mm, (b) Scale = _____ ″/mm

4. (a) _____ mm, (b) Parallax angle $\theta =$ _____ ″

5. $r =$ _____ km

6. (a) 1 au = _____ km

   (b)

7. (a)

(b) $r =$ _____ km

(c)

(d)

8.

Introduction to Stars and Planets
An activities-based exploration
Alan Hirshfeld

Activity 3

The Sun's Diameter and Mass

Preview

The Sun's diameter is inferred from its measured distance and angular span in the sky. Its mass is computed using Kepler's third law of planetary motion, a mathematical formula derived during the 1600s.

3.1 The Sun's Diameter

In the previous activity, we explored observational techniques by which the Sun's distance can be determined. Once that distance is known, we are able to compute a pair of other key physical attributes of the Sun: its diameter and its mass. The Sun's *angular diameter*—the angle it spans in the sky—is measurable with simple instruments. But the Sun's angular span alone reveals nothing about its true physical extent, or *linear diameter*, in, say, units of kilometers: we cannot tell whether the Sun is a relatively compact fireball somewhere close to our planet or a huge luminary many millions of kilometers away. To compute the Sun's actual size, we need to take account of its distance from the Earth.

The route to measurement of the Sun's linear diameter begins with a standard geometric formula that relates the angular span of a sector of a circle to the length of the arc enclosed by that sector. (For example, a slice of apple pie forms a sector, whose curved segment of crust is its enclosed arc length.) Figure 3.1 illustrates a sector: the Greek letter θ represents the angular width of a sector of a circle; s is the length of the arc enclosed by that sector; and r is the circle's radius. These quantities are mathematically related by the sector formula:

$$s = (r \times \theta)/57.3, \tag{3.1}$$

where r and s are both expressed in the same unit of length and θ is expressed in degrees. (An alternative version of this formula in Activity 2 assumed that θ is given in units of arcseconds instead of degrees; hence, the difference in the numerical constant.)

doi:10.1088/2514-3433/abc249ch3   © IOP Publishing Ltd 2020

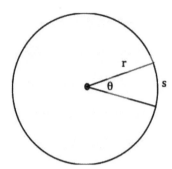

Figure 3.1. Sector of a circle of radius *r*. The sector encloses an angle *θ* and an arc of length *s*.

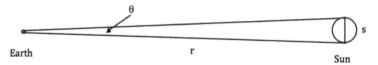

Figure 3.2. A sector-like representation of the Earth and the Sun. The sector, centered on the Earth, encloses the arc length *s*, which is a close approximation to the Sun's linear diameter.

Figure 3.2 depicts the sector-like arrangement of the Earth–Sun system, with *s* now representing the Sun's diameter and *r* the Sun's distance from the Earth. Note: Figure 3.2 is not drawn to scale; indeed, the Earth is so far from the Sun that the arc length *s* is indistinguishable from the Sun's actual diameter.

We can use Equation (3.1) to deduce the Sun's linear diameter, since both its angular diameter in the sky *θ* and its distance *r* are measurable. (The same can be said for the Moon, a planet, or any other celestial object with a measured angular diameter and distance.) The Sun's angular span in the sky is **0.52°** and its average distance from Earth is 1 astronomical unit (au), or roughly **150 million kilometers (km)**.

1. Use Equation (3.1) to compute the Sun's true diameter *s*, in kilometers (km).
2. Given that the Earth's diameter is about 12,800 km, approximately how many Earths would fit alongside each other across the diameter of the Sun?

3.2 The Sun's Mass

The Sun's mass can be determined from Kepler's third law of planetary motion, as refined by Isaac Newton in the late 1600s. This law expresses the mathematical relationship among three quantities: (i) a planet's orbital period *P*, which is the time it takes the planet to orbit the Sun; (ii) the radius *r* of the planet's orbit, which we will assume, for simplicity's sake, is a circle; and (iii) the respective masses of the planet and the Sun. Newton's version of Kepler's third law is written as follows:

$$P^2 = \frac{4\pi^2 r^3}{G(M_s + M_E)},$$ (3.2)

where G is the universal gravitational constant (which has the numerical value 6.667×10^{-11}), M_S is the mass of the Sun, and M_E is the mass of the Earth. Simply stated, Kepler's third law allows us to use the Earth's own orbital period and orbital radius to compute the Sun's mass! We can safely assume that the Earth's mass is numerically insignificant compared to that of the Sun; in other words, the sum $M_S + M_E$ in the denominator (bottom) of the right-hand side of Equation (3.2) is very closely approximated by M_S alone:

$$P^2 \approx \frac{4\pi^2 r^3}{GM_s}. \tag{3.3}$$

Here the symbol "≈" means "approximately equal to." Multiplying both sides of Equation (3.3) by M_S, then dividing both sides by P^2 yields the following expression for the Sun's mass M_S:

$$M_S \approx \frac{4\pi^2 r^3}{GP^2}. \tag{3.4}$$

3. In order for the units in Equation (3.4) to combine correctly, convert the radius r of the Earth's orbit—approximately 150 million kilometers $(1.5 \times 10^8 \text{ km})$—into units of *meters*. There are 1000 m (10^3 m) in a kilometer. Express your answer in powers-of-ten notation, keeping two decimal places in the leading number. For guidance, refer to the mathematics tutorial in the Appendix.

4. In Equation (3.4), we must also convert the Earth's orbital period P—1 yr—into units of seconds. Knowing the number of days in a year, the number of hours in a day, the number of minutes in an hour, and the number of seconds in a minute, convert 1 yr into days, then into hours, then into minutes, and finally into seconds. Express your answer in powers-of-ten notation, keeping two decimal places in the leading number.

5. Now use Equation (3.4) to compute the mass of the Sun, which will come out in units of kilograms (kg). Once again, express your answer in powers-of-ten notation, keeping two decimal places in the leading number.

6. Scientists have estimated the Earth's mass at 6×10^{24} kg. How many times greater is the Sun's mass than the Earth's mass? That is, compute the ratio of the Sun's mass to the Earth's mass. Obey the following rule: when dividing two quantities in powers-of-ten notation, divide the leading numbers ("coefficients") and subtract the powers ("exponents") from one another.

7. Virtually all of the solar system's *planetary* mass lies in the so-called "gas giants" Jupiter, Saturn, Uranus, and Neptune. Together, these weighty worlds add up to the mass equivalent of some 450 Earths. (a) How many times greater is the Sun's mass than the total mass of its planets? (b) From your answer to part 7(a), are the planets a major or a minor contributor to the overall mass of the solar system, including the Sun?

3.3 Implications

Having determined the Sun's diameter, and recognizing the obvious fact that it is spherical, a standard geometrical formula allows us to compute the Sun's overall volume. The answer is an enormous number, perhaps best conveyed by the conclusion that about 1 million Earths could fit inside the Sun. And with the Sun's mass likewise known, it's possible to deduce the Sun's average density: the average concentration of matter within it, or technically its mass per unit volume. This number comes out relatively small—about 1 gram per cubic centimeter—close to the density of water and much less dense than a rocky planet like our Earth. We infer that the Sun consists mostly of lightweight gases, such as hydrogen and helium, highly compressed in the Sun's core, but in a near-vacuum condition at its surface.

Worksheet, Activity 3: The Sun's Diameter and Mass

Name _____

1. Sun's diameter = _____ km

2. Sun's diameter = _____ Earth-diameters

3. 1 au = _____ m

4. 1 yr = _____ s

5. Sun's mass = _____ kg

6. Sun's mass = _____ Earth-masses

7. (a) Sun's mass = _____ times combined planetary mass

   (b) _____

Introduction to Stars and Planets
An activities-based exploration
Alan Hirshfeld

Activity 4

The Sunspot Cycle

Preview

The occurrence of sunspots has been recorded by solar observers for centuries. A graph of their numbers reveals a cyclical pattern in time.

4.1 Tracking Sunspots Through History

Sunspots are dark patches on the solar surface that last from weeks to months before dissipating. (Several sunspots are visible in Figure 2.1 of Activity 2.) At around 4500°C, sunspots are typically about a thousand degrees cooler than their surroundings, accounting for their apparent darkness; absent the brilliant solar backdrop, sunspots would glow like any other hot, incandescent gas. Sunspots are centers of intense magnetic activity, which redirects the outflow of solar energy to adjacent surface regions in the form of solar flares, prominences, and mass ejections.

For centuries, astronomers have kept track of the number and positions of sunspots. They examine this data for variations—especially cyclical variations—in both the annual sunspot count and the solar latitude at which spots occur. Table 4.1 lists the officially recognized yearly average sunspot number from 1700 through 2015, obtained by taking the average of the daily counts of sunspots for a given year. As evidenced by the data in the table, during times of peak activity, hundreds of spots might be visible on the solar surface!

1. In Table 4.1, circle the years in which the annual sunspot count achieved a *maximum*. For example, the year 1705 reached a count of 97 spots, with fewer spots during the preceding and subsequent years. For consecutive or nearly consecutive years of high sunspot numbers, circle only the year with the largest count. Enter all of the circled years and their corresponding sunspot numbers in Table 4.2 on the worksheet; the year and sunspot number for the first two maxima are shown.

doi:10.1088/2514-3433/abc249ch4
© IOP Publishing Ltd 2020

Table 4.1. Sunspot Numbers from 1700 to 2015

Year	No.	Year	No.	Year	No.	Year	No.	Year	No.	Year	No.	Year	No.	Year	No.
1700	8	1741	67	1782	64	1823	2	1864	89	1905	106	1946	154	1987	34
1701	18	1742	33	1783	38	1824	11	1865	58	1906	90	1947	215	1988	123
1702	27	1743	27	1784	17	1825	28	1866	31	1907	103	1948	193	1989	211
1703	38	1744	8	1785	40	1826	60	1867	14	1908	81	1949	191	1990	192
1704	60	1745	18	1786	138	1827	83	1868	63	1909	73	1950	119	1991	203
1705	97	1746	37	1787	220	1828	109	1869	124	1910	31	1951	98	1992	133
1706	48	1747	67	1788	218	1829	115	1870	232	1911	10	1952	45	1993	76
1707	33	1748	100	1789	197	1830	117	1871	185	1912	6	1953	20	1994	45
1708	17	1749	135	1790	150	1831	81	1872	169	1913	2	1954	7	1995	25
1709	13	1750	139	1791	111	1832	44	1873	110	1914	16	1955	54	1996	12
1710	5	1751	80	1792	100	1833	13	1874	75	1915	79	1956	201	1997	29
1711	0	1752	80	1793	78	1834	20	1875	28	1916	95	1957	269	1998	88
1712	0	1753	51	1794	68	1835	86	1876	19	1917	174	1958	262	1999	136
1713	3	1754	20	1795	36	1836	193	1877	21	1918	135	1959	225	2000	174
1714	18	1755	16	1796	27	1837	227	1878	6	1919	106	1960	159	2001	170
1715	45	1756	17	1797	11	1838	169	1879	10	1920	63	1961	76	2002	164
1716	78	1757	54	1798	7	1839	143	1880	54	1921	44	1962	53	2003	99
1717	105	1758	79	1799	11	1840	106	1881	91	1922	24	1963	40	2004	65
1718	100	1759	90	1800	24	1841	63	1882	99	1923	10	1964	15	2005	46
1719	65	1760	105	1801	57	1842	40	1883	106	1924	28	1965	22	2006	25
1720	47	1761	143	1802	75	1843	18	1884	106	1925	74	1966	67	2007	13
1721	43	1762	102	1803	72	1844	25	1885	86	1926	107	1967	133	2008	4
1722	37	1763	75	1804	79	1845	66	1886	42	1927	115	1968	150	2009	5
1723	18	1764	61	1805	70	1846	103	1887	22	1928	130	1969	149	2010	25
1724	35	1765	35	1806	47	1847	166	1888	11	1929	108	1970	148	2011	81
1725	67	1766	19	1807	17	1848	208	1889	10	1930	59	1971	94	2012	85
1726	130	1767	63	1808	14	1849	183	1890	12	1931	35	1972	98	2013	94
1727	203	1768	116	1809	4	1850	126	1891	60	1932	19	1973	54	2014	113
1728	172	1769	177	1810	0	1851	122	1892	122	1933	9	1974	49	2015	82
1729	122	1770	168	1811	2	1852	103	1893	142	1934	15	1975	23		
1730	78	1771	136	1812	8	1853	74	1894	130	1935	60	1976	18		
1731	58	1772	111	1813	20	1854	39	1895	107	1936	133	1977	39		
1732	18	1773	58	1814	23	1855	13	1896	69	1937	191	1978	131		
1733	8	1774	51	1815	59	1856	8	1897	44	1938	183	1979	220		
1734	27	1775	12	1816	76	1857	43	1898	44	1939	148	1980	219		
1735	57	1776	33	1817	68	1858	104	1899	20	1940	113	1981	199		
1736	117	1777	154	1818	53	1859	178	1900	16	1941	79	1982	162		
1737	135	1778	257	1819	39	1860	182	1901	5	1942	51	1983	91		
1738	185	1779	210	1820	24	1861	147	1902	9	1943	27	1984	61		
1739	168	1780	141	1821	9	1862	112	1903	41	1944	16	1985	21		
1740	122	1781	114	1822	6	1863	84	1904	70	1945	55	1986	15		

2. In Table 4.2, enter the number of years **Y** between each successive pair of sunspot maxima. An example is shown: the number of years between the first two sunspot maxima is $1717-1705 = 12$ years.

3. Compute the average of all the **Y** time intervals you listed in Table 4.2. Enter your result on the worksheet. This is the average period of the sunspot cycle.

4. From your data in Table 4.2, compute the average of the maximum sunspot numbers from 1700 through 2015. That is, sum up all the numerical entries in the "Number" columns, then divide by the total number of entries.

5. Now underline all the years in which the annual sunspot count was at a minimum *between the years 1900 and 2000 only*. You will need this data for the next part.

6. Sometimes it's hard to identify a cyclical pattern in a tabular collection of data, as is the case here for sunspots numbers. But a graphical representation of the data often reveals an otherwise hidden pattern.

   (a) On the axes provided in Figure 4.1 on the worksheet, graph the sunspot number for each year of sunspot maximum from 1900 to 2000. On the same axes, graph the sunspot number for each year of sunspot minimum from 1900 to 2000 (see part 5 above).

   (b) On your graph, draw a line to connecting the first sunspot maximum data point to its subsequent sunspot minimum, then from that sunspot minimum to its subsequent maximum. Repeat the process until all of the maximum–minimum data points are connected, forming a series of alternating spikes and valleys.

   (c) Draw a horizontal line on the graph representing the historical average maximum sunspot number that you computed in part 4.

   (d) On the graph, indicate any years or sequence of years when the maximum sunspot number was either significantly higher than average or significantly lower than average. During one such period, known as the "Maunder Minimum," from 1645 to about 1700, sunspots were virtually absent from the Sun's face! To date, there is no generally accepted explanation for this occurrence.

Worksheet, Activity 4: The Sunspot Cycle

Name _____

1, 2. Sunspot maxima.

Table 4.2. Fill-in Table of Sunspot Maxima

Year	Number	Y	Year	Number	Y	Year	Number	Y
1705	97							
1717	105	12						

3. Average of **Y** intervals (1700–2015) = _____ yr

4. Average of maximum sunspot numbers (1700–2015) = _____

5. Sunspot minima (see Table 4.1).

6. (a), (b), (c), (d)

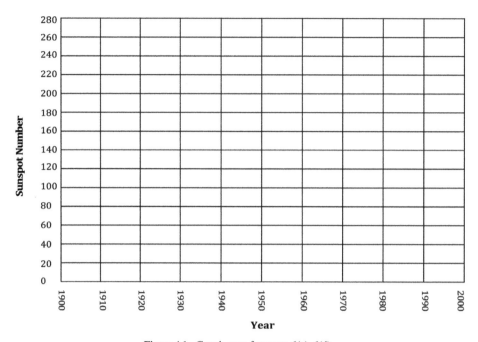

Figure 4.1. Graph axes for parts 6(a)–6(d).

Introduction to Stars and Planets
An activities-based exploration
Alan Hirshfeld

Activity 5

The Solar Constant

"The Sun's rays are the ultimate source of almost every motion which takes place on the surface of the Earth. By its heat are produced all winds ... By their vivifying action, vegetables are elaborated from inorganic matter, and become, in their turn, the support of animals and of man, and the sources of those great deposits of dynamical efficiency which are laid up for human use in our coal strata."

—John Herschel, Treatise on Astronomy, 1833.

Preview

Measurements of solar energy at various wavelengths reaching the Earth are used to derive a key astronomical parameter called the *solar constant*. Determination of this constant is necessary to calculate the overall amount of energy emitted by the Sun: its *luminosity*.

5.1 Basic Concepts

The Sun has been a steady source of energy for the whole of human history—and, as we will later learn, for far longer than that. The influx of solar photons sustains an array of natural processes, many of which are essential to the maintenance of life on our planet. Among the solar properties of intense interest to astronomers is the Sun's energy output. We cannot travel to the Sun's surface to measure this luminous outflow directly; no spacecraft would survive such a mission. Instead, we deduce the Sun's overall emissive power in a two-step process: first by measuring the amount of solar energy that reaches the Earth: the solar constant; then by applying a geometry-based adjustment to this number to infer the Sun's overall luminosity.

1. The intensity of sunlight has been carefully measured since the mid-19th century by detectors on the ground and, more recently, by detectors on satellites that orbit our Earth. Why do you think measurements of the solar constant made by detectors in space are much preferred to those made by detectors on Earth's surface? Be specific.

2. To correctly gauge the solar constant, a satellite must be positioned as far away from the Sun as is our planet, that is, 1 astronomical unit (au) from the Sun. A satellite circling Earth at an altitude of, say, 200 km might pass as much as 200 km closer to the Sun than the ground below, depending on the orbit's orientation. Given that Earth's average distance from the Sun is about 150 million kilometers, do you think a variation in distance of 200 km matters to a measurement of the solar intensity? Support your answer by computing the fraction 200 km over 150 million kilometers.

5.2 Measuring the Solar Constant

Figure 5.1 is a graph depicting the measured distribution of solar energy above Earth's atmosphere over a wide range of wavelengths, an array of energy called the *solar spectrum*. The figure's longer axis indicates the wavelength of energy in units of nanometers. (A nanometer, abbreviated nm, is a billionth of a meter: 10^{-9} m in powers-of-ten notation.) Wavelengths visible to the human eye, which together comprise the *optical* portion of the Sun's spectrum, fall within the range of about 400 nm (violet) to 700 nm (red).

3. On the wavelength axis of Figure 5.1, draw marks to indicate the extremes of the optical portion of the Sun's spectrum. If the eventual goal is to compute the Sun's *overall* luminosity, why is it necessary to measure the Sun's emission at both optical and non-optical wavelengths?

The shorter axis of Figure 5.1 displays the intensity of the Sun's light expressed in units of watts per square meter per nanometer, abbreviated $W\ m^{-2}\ nm^{-1}$. There are three components to this rather cumbersome unit:

- The number of watts (W) indicates the rate at which energy is emitted, whether for the Sun or for a household light bulb. That is, a 100 W light bulb will emit more energy each second than a 60 W bulb, and will therefore appear brighter when viewed from the same distance.
- The Sun's emission shines in all directions. However, let's focus exclusively on the solar wattage shining on a make-believe light sensor whose shape is a square, 1 meter on a side. That's what the term "per square meter" (m^{-2}) refers to in the light-intensity unit of Figure 5.1. In effect, we're sampling only the sunlight that shines onto our imagined $1\ m^2$ detector and ignoring the rest.
- The third part of the light-intensity unit, "per nanometer" (nm^{-1}), indicates that the numbers along this axis of Figure 5.1 refer to the sunlight intensity measured within very narrow wavelength segments, each just 1 nm wide.

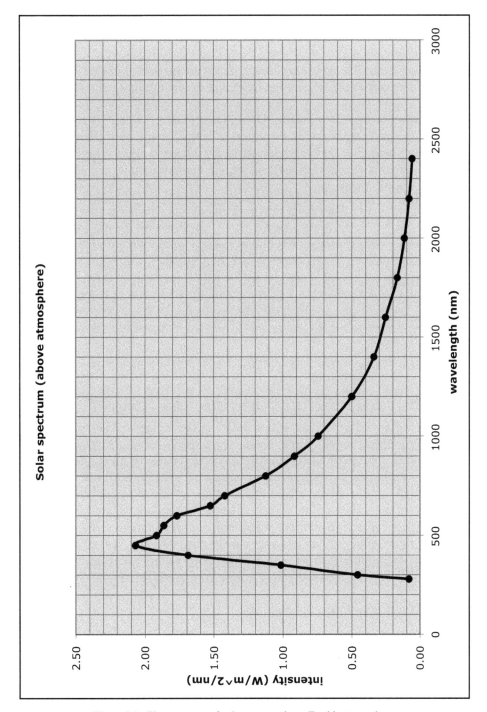

Figure 5.1. The spectrum of solar energy above Earth's atmosphere.

For example, the Sun gives out less energy at a wavelength of 300 nm (ultraviolet) than at 700 nm (red): about 0.45 W m^{-2} nm^{-1} compared to 1.4 W m^{-2} nm^{-1}.

To determine the solar constant *S*—the overall intensity of sunlight illuminating a 1 meter-wide square at the Earth—we must sum up the amounts of energy received at the various individual wavelengths. This is easily done using Figure 5.1. In fact, the *area* of the graph underneath the solar intensity curve represents precisely what we are seeking: the solar constant *S*, in units of watts per square meter (W m^{-2}). To ease the determination of the area underneath the intensity curve, the graph has been divided into small boxes. Counting these boxes is the first step to achieve a value for the area we seek to measure.

4. Count the number of boxes underneath the solar intensity curve in Figure 5.1. For boxes that are divided by the curve itself, adopt the following procedure: estimate what fraction of the box lies underneath the curve. If the fraction underneath the curve is less than one-half, omit that box from the count; if the fraction is greater than one-half, include the box in the count.

5. So far, all we have determined is the *number* of boxes that lie underneath the solar intensity curve. However, the *area* of each box represents a specific amount of solar energy that can be estimated from the scale units of the axes of Figure 5.1.
   (a) What is the width of each box, in *nanometers*, according to the graph's wavelength scale?
   (b) What is the height of each box, in watts per square meter per nanometer, according to the graph's intensity scale?
   (c) The area of a rectangle is simply its width times its height. Therefore, multiply together your answers to parts 5(a) and 5(b) to yield the area of one box, in watts per square meter. That is, each box in your count from part 4 represents this amount of solar energy.

6. Multiply your answer to part 5(c) by your count of boxes underneath the solar intensity curve from part 4. The resulting number is your estimation of the solar constant *S*, the overall amount of solar energy at all wavelengths that illuminates a 1 meter-wide square each second at the Earth. Quick check: the standard value of *S* is about 1370 W m^{-2}; if your answer differs substantially from this number, review your work for errors.

5.3 The Solar Constant in Practice

The Sun is quite generous with its luminous energy, imparting nearly 1400 W of power onto a 1 m^2 surface above the Earth's atmosphere. However, harnessing this solar energy for practical use presents some challenges. While solar illumination might be a convenient source of supply to meet the energy needs of an orbiting space station, capturing solar rays on the Earth's surface for commercial or domestic purposes requires adjustment of the "raw" solar constant calculated in the previous part. For example, assuming the sky is cloudless, only about 30% of sunlight hitting the uppermost layer of the Earth's atmosphere completes the downward journey to sea level. Furthermore, at present, the typical household solar panel is only 20% efficient, that is, only 20%, or one-fifth, of the sunlight shining on the panel is converted into electricity.

7. (a) Revise your value of the solar constant S to account for the diluting effect of the Earth's atmosphere and for the low efficiency of present-day solar panels, as described above. (b) What other factors must be considered when implementing a ground-based, solar electricity-generation system? Explain each.

8. Despite the practical obstacles, solar energy is an attractive supplemental source of electrical power, given the steady stream of sunlight that irradiates the Earth's daytime hemisphere. Let's compute the total amount of luminous energy intercepted by the Earth. To simplify the calculation, we "flatten" our planetary sphere into a circular disk of identical radius. This Earth-sized disk is oriented face-on to the solar rays. Its area is given by the formula $A = \pi R^2$, where R is the Earth's radius, 6.378×10^6 m.

   (a) Compute the area A of an Earth-size disk, in square meters.

   (b) To find the overall solar illumination on Earth, in watts, multiply the area A by the accepted value of the solar constant S (1370 W m^{-2}).

   (c) The world's present-day rate of energy consumption is around 1.3×10^{13} W. Compute the ratio of the overall solar illumination on our planet, from part 8(b), to the rate at which humanity currently consumes energy.

   (d) Express in your own words the meaning of the ratio you computed in part 8(c).

Worksheet, Activity 5: The Solar Constant

Name _____

1. _____
   _____
   _____

2. _____
   _____
   _____

3. _____
   _____

4. Number of boxes = _____

5. (a) Width of box = _____ nm
   (b) Height of box = _____ $W\ m^{-2}\ nm^{-1}$
   (c) Area of box = _____ $W\ m^{-2}$

6. Solar constant S = _____ $W\ m^{-2}$

7. (a) "Revised" solar constant = _____ $W\ m^{-2}$

   (b) _____
       _____
       _____
       _____
       _____
       _____

8. (a) A = _____ m^2

   (b) Solar illumination on Earth = _____ W

   (c) Ratio = _____

   (d) _____
       _____

Introduction to Stars and Planets
An activities-based exploration
Alan Hirshfeld

Activity 6

The Sun's Luminosity

Preview

The solar constant—the solar energy illuminating a 1 m^2 detector at the Earth per second—is used to calculate the Sun's luminosity: the total amount of energy emitted by the Sun per second into outer space.

6.1 Introduction

One of the key attributes of the Sun, besides its size, mass, temperature, and chemical composition, is its *luminosity L*: the total amount of energy the Sun emits each second in all directions. This energy comprises a broad spectrum of wavelengths from the ultraviolet through the visible and into the infrared. There are also lesser amounts of X-rays and radio waves. In the previous activity, we took the first step to gauge the Sun's overall energy output: we computed the solar constant *S*, a measure of the solar energy reaching the Earth. But the Sun radiates equally in all directions, rendering the solar constant a minuscule fraction of our star's overall energy outflow, as illustrated in Figure 6.1. Nevertheless, geometry provides us with a simple path to transform the value of the solar constant into the total amount of energy emitted by the Sun.

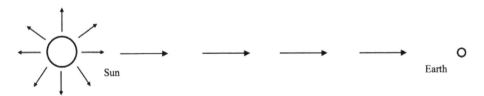

Figure 6.1. Earth intercepts only a tiny fraction of the Sun's energy. (Figure not to scale.)

doi:10.1088/2514-3433/abc249ch6

6-1

© IOP Publishing Ltd 2020

6.2 Squares and Spheres

By definition, the solar constant S is the amount of solar energy striking a 1 meter-wide square each second at the Earth's distance from the Sun. Imagine placing a second 1 m^2 alongside this first square; together these squares intercept twice the energy of a single square, that is, $2 \times S$. Adding a third 1 m^2 raises the collective solar energy intercepted to $3 \times S$. For 4 such squares, the total solar energy is $4 \times S$, and so forth.

Now imagine assembling an immense number of these squares, curving them gently so as to construct an enormous sphere that completely encloses the central Sun, as depicted in Figure 6.2. Any light energy leaving the Sun's surface must necessarily pass through this sphere on its way into deep space. (Let's ignore the tiny fraction of solar energy intercepted by the solar system's inner planets Mercury and Venus.)

To find the Sun's luminosity L, we simply multiply the solar constant S by the surface area A of this giant sphere, whose radius is equal to the radius of the Earth's orbit, 1 au. (Don't confuse the radius of our planet with the radius of its orbit, which is more than 20,000 times larger.)

The surface area A of a sphere is given by the formula:

$$A = 4\pi R^2, \tag{6.1}$$

where R is the sphere's radius. In this case, R is the average radius of the Earth's orbit: 1 au, or 1.50×10^{11} m. (This particular application requires us to express the au in units of meters instead of the usual kilometers.)

1. Using the surface-area formula, Equation (6.1), compute the area A of the Sun-enclosing sphere described above.
2. To find the Sun's luminosity L, in watts (W), multiply the value of the solar constant S (1370 W m^{-2}) by your answer for the area A in part 1. A watt is a standard measure of energy production or usage, such as a 100 W light bulb or a 1000 W space heater. Quick check: the accepted value of L is about 3.84×10^{26} W; if your answer differs substantially from this standard value, review your work for errors.

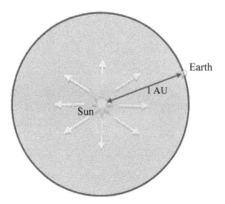

Figure 6.2. Solar energy passing through an imaginary sphere concentric with the Sun.

As far as stellar luminosity goes, the Sun is no champion. Sirius, the brightest star in the night sky, has an energy emission rate some 25 times that of our Sun, whereas the red supergiant star Betelgeuse in the constellation Orion spews the energy equivalent of ten-thousand Suns. There exist rare stars whose luminosity, at times, exceeds a million Suns! Although super-luminaries like Sirius and Betelgeuse had long been known, it was not until the 20th century that physicists determined precisely *how* the Sun and its stellar siblings generate their prodigious amounts of energy.

Worksheet, Activity 6: The Sun's Luminosity

Name _____

1. Surface area A of Sun-enclosing sphere = _____ m^2

2. Sun's luminosity L = _____ W

Introduction to Stars and Planets
An activities-based exploration
Alan Hirshfeld

Activity 7

The Sun's Surface Temperature

Preview

The Sun's surface temperature is determined from physical principles involving the surface area and energy output of hot, incandescent objects. The same principles allow comparison of the Sun's radiant emission and size to those of other stars.

7.1 Taking the Sun's Temperature

In a previous activity, we determined the Sun's true diameter from its angular diameter and distance. Next we measured the solar constant and used it to derive the Sun's overall energy output, or luminosity. From these results, we can now determine another important physical attribute of the Sun: its surface temperature. Hot, luminous objects like the Sun—say, a glowing light bulb filament or a bucket of molten steel—approximate the physicist's idealized energy-emitter known as a *blackbody*; as such, their emissive properties conform, more or less, to a specific mathematical rule. For a hot, incandescent sphere, this rule can be expressed as a formula that relates the sphere's luminosity L, radius R, and surface temperature T:

$$L = (4\pi R^2) \times (\sigma T^4). \qquad (7.1)$$

Here, the luminosity L is expressed in watts (W)—the same unit used to quantify the brightness of a light bulb—and the radius R is expressed in units of meters (m). The object's surface temperature T is measured on the Kelvin (K) scale, which is commonly used in the physical sciences. The Kelvin system effectively shifts the zero point of the familiar centigrade (°C) system to *absolute zero*, the lowest achievable temperature, according to physics theory; thus, absolute zero is about −273 °C. In Kelvin units, the freezing point of water (0°C) is approximately 273 K, while the boiling point (100°C) is 373 K. The Greek letter σ (sigma) in Equation (7.1) stands for a fundamental number in physics called the Stefan–Boltzmann constant.

doi:10.1088/2514-3433/abc249ch7

© IOP Publishing Ltd 2020

The most remarkable feature of Equation (7.1) is the large numerical exponent applied to the temperature **T**. The exponent predicts that if the temperature of an incandescent object is, for instance, doubled, the object's luminosity escalates by a factor of 2^4, or 16 times. In other words, a star the size of our Sun, but with twice the Sun's surface temperature, emits 16 times more energy into space!

Since we previously determined the Sun's luminosity **L** (3.84×10^{26} W) and radius **R** (one-half its diameter, or 6.96×10^8 m), we can use Equation (7.1) to reveal the Sun's surface temperature **T** without having to immerse a rocket-borne, high-tech thermometer into the Sun itself. To simplify the computation, let's plug in the numerical value of the fundamental constant σ and then solve Equation (7.1) for **T**:

$$T = 34 \times \sqrt[4]{\frac{L}{R^2}}. \tag{7.2}$$

Note: The fourth root of a number, represented by the symbol $\sqrt[4]{}$, is equivalent to applying your calculator's square root function twice in succession, that is, taking the square root of the square root.

1. Use Equation (7.2) to compute the Sun's surface temperature in Kelvin units.
2. Convert your answer to part 1 into centigrade degrees by subtracting 273 from the Kelvin temperature. Quick check: the accepted value of the Sun's surface temperature is about 5500 °C; if your answer differs substantially from this value, review your work for errors.
3. Consult sources on the Internet or in reference books to compare the Sun's surface temperature to those of incandescent substances on the Earth, such as volcanic lava, an oxyacetylene torch, or molten tungsten, which has the highest melting point of any chemical element. (By the way, the laws of physics indicate that the Sun's internal temperature is far higher than that of its surface, rising to some 15 million degrees at the center!)
4. The blue supergiant star Rigel, in the constellation of Orion, has a luminosity about 120,000 times that of the Sun and a radius of about 75 times the Sun's. Use Equation (7.2) to compute Rigel's surface temperature in both Kelvin and centigrade units.

7.2 A Pint-sized Star

With a bit of algebra, Equation (7.2) can be rewritten to yield a formula for the radius of a star, in units of meters:

$$R = \sqrt{\frac{L}{(T/34)^4}}. \tag{7.3}$$

5. The star Sirius, in the constellation Canis Major, has a small companion star—a *white dwarf*—whose luminosity is a paltry one-fortieth (or 0.025) times that of the Sun and whose surface temperature is about 25,000 K.

(a) Use Equation (7.3) to compute the radius of this white dwarf star in both meters and kilometers. (There are 1000 m in a kilometer.)

(b) Compare the radius of the white dwarf to the radius of the Sun, that is, divide the white dwarf's radius by the Sun's radius, both expressed in kilometers. Comment on the meaning of your numerical result.

(c) Similarly, compare the radius of the white dwarf to the Earth's radius, about 6371 km. Comment on the meaning of your numerical result.

(d) From the undulating space-movement of the white dwarf relative to its stellar partner, Sirius, we can infer that the white dwarf "weighs" about as much as the Sun. Considering your answers to parts 5(b) and 5(c), what can you deduce about the average density, or concentration of matter, of the white dwarf compared to the density of everyday objects? Explain the reasoning behind your answer.

Worksheet, Activity 7: The Sun's Surface Temperature

Name _____

1. $T =$ _____ K

2. $T =$ _____ °C

3. _____
   _____
   _____

4. $T =$ _____ K = _____ °C

5. (a) $R =$ _____ m = _____ km

   (b) White dwarf's radius/Sun's radius = _____
   _____
   _____

   (c) White dwarf's radius/Earth's radius = _____
   _____
   _____

   (d) _____
   _____
   _____

Introduction to Stars and Planets
An activities-based exploration
Alan Hirshfeld

Activity 8

Spectral Lines and the Chemistry of the Sun

Preview

Analysis of features in the spectrum of the Sun's light reveals the chemical constituents of its atmosphere.

8.1 Dark Lines and Bright Lines

In 1814, the German optical craftsman Joseph von Fraunhofer studied the Sun's spectrum through a precision spectroscope that he built in his workshop. Interspersed among the expected rainbow colors of the spectrum were hundreds of fine, dark lines—apparent absences of color—to which he assigned a series of uppercase and lowercase letters. Neither he nor anyone else at the time understood the origin or significance of these "Fraunhofer lines," as they came to be called. In the decades that followed, laboratory chemists found similar line patterns in the spectra of flames seeded with various atoms, molecules, or compounds, except that these lines were bright instead of dark.

By 1859, physicist Gustav Kirchhoff and chemist Robert Bunsen, at Heidelberg University, in Germany, realized that the patterns of dark lines in the Sun's spectrum coincided with the patterns of bright lines emitted by burning substances in the laboratory; hence, by pattern matching, spectral lines could be used to deduce the presence of specific elements in the atmospheres of the Sun and the stars. In a star's atmosphere, Kirchhoff explained, an element's atoms *absorb* specific colors (wavelengths) of light before they escape into outer space, creating absences of color —dark lines—within the star's spectrum. Atoms of that same element, when ignited in a laboratory flame, *emit* an identical array of spectral lines, only bright instead of dark, as shown in Figure 8.1.

It was not until the early 20th century that advances in atomic and radiative physics revealed how the width and the relative opacity (darkness) of the Fraunhofer lines reflect the comparative proportions of chemical elements in a stellar atmosphere. Nevertheless, by the late 1800s, the remoteness of stars no longer prevented

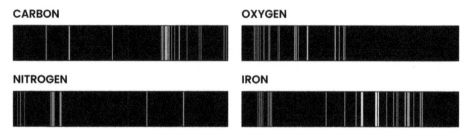

Figure 8.1. Bright-line spectra of the elements carbon, oxygen, nitrogen, and iron, as they would appear in a laboratory spectroscope. Their dark-line counterparts are found in the Sun's spectrum, indicating the presence of these elements in the solar atmosphere. (Credit: NASA/STScI.)

the study of their elemental makeup; the chemical constitution of these distant suns is conveyed across the vast gulf of space, encoded in their light.

8.2 The Sun's Spectrum

Figure 8.2 on the worksheet depicts a black-and-white rendering of the Sun's visible spectrum. Were the spectrum presented in color, short-wavelength violet light would lie at the left end of the figure and long-wavelength red light would lie at the right end, with the other rainbow hues in between. Indeed, a color representation is unnecessary, since we will be using a numerical proxy—wavelength—to situate the various features in the spectrum.

Given the wavelengths of two of the Fraunhofer lines, designated C and F in Figure 8.2, we will first determine (i) the wavelength scale of the spectrum image and (ii) the wavelength of each of the other Fraunhofer lines in the figure. Then we will consult a reference list of Fraunhofer lines to find the alphabetical letter Fraunhofer had assigned to each line, as well as the chemical element whose light-absorption created it. (There is no relation between Fraunhofer's letters and the letter abbreviations representing chemical elements.) For example, the C and F lines are both caused by the absorption of solar photons by hydrogen atoms in the Sun's atmosphere: the C line has a wavelength of 656 nm and the F line 486 nm. (A nanometer, abbreviated nm, is one-billionth of a meter; visible light spans the wavelength range from about 400 nm to about 700 nm.)

8.3 Establishing the Spectrum's Wavelength Scale

The reproduction of the solar spectrum in Figure 8.2 is analogous to a geographic map: the spectrum represents a scaled version of the original spectrum, just as the map represents a scaled rendering of a region on the Earth. Whether displayed on a computer screen, reproduced in a book, or printed out on a piece of paper, the scale of the spectrum must be determined; that is, how many nanometers in wavelength is represented by a given millimeter measurement in Figure 8.2? To determine the spectrum's scale, we need to know the actual wavelengths, in nanometers, of two spectral lines, here, the C and F lines.

1. (a) Use a metric ruler to measure the separation **Δd** (pronounced "delta d"), in millimeters, between the C line and the F line in Figure 8.2. Measure to the nearest *tenth* of a millimeter, that is, your measurement should be expressed to one decimal place.
   (b) From the data given above, compute the wavelength difference **Δw** ("delta w"), in nanometers, between the C line and the F line.
   (c) Divide the wavelength difference **Δw** of these two lines by their separation **Δd** to yield the scale *s* of the spectrum image, in units of nanometers per millimeter, or nm/mm. In other words, each measured millimeter in Figure 8.2 corresponds to *s* nanometers in wavelength. Enter your answers on the worksheet.

8.4 Computing the Wavelengths of the Dark Lines

2. Starting from the left (violet) end of Figure 8.2, determine the wavelengths of the dark lines labeled 1 through 9. The method is simple:
   (i) Measure the number of millimeters **Δmm** between the selected line and either the C or the F line—your chosen "reference line."
   (ii) Use the image scale *s* you derived in part 1 to convert the millimeter-difference **Δmm** on the spectrum photo into a nanometer-difference **Δnm**.
   (iii) To compute the wavelength *w* of your selected line:
   - If the selected line is to the right of your chosen reference line (toward longer wavelength), *add* the nanometer-difference **Δnm** to the wavelength of the reference line.
   - If the selected line is to the left of your chosen reference line (toward shorter wavelength), *subtract* the nanometer-difference **Δnm** from the wavelength of the reference line.

Write your answers in Table 8.2 on the worksheet.

8.5 Identifying the Fraunhofer Lines

3. Table 8.1 is a partial list of the Sun's Fraunhofer lines. Compare your wavelength determinations for lines 1–9 in the previous part to the wavelengths in the list. If you find a match or near match (values within, say, 5–10 nm of each other), write the corresponding letter designation and its wavelength in Table 8.2 on the worksheet.

Upon completion, you will have successfully performed a sunlight-based, chemical analysis of the Sun! The same spectroscopic process applies to stars beyond our solar system, as well as to entire galaxies of stars outside our Milky Way galaxy. Spectrum analysis of stars at the utmost reach of our telescopes reveals that far-flung celestial objects are composed of the same array of chemical elements that constitute matter here on Earth and in the near-universe.

Table 8.1. Prominent Fraunhofer Lines in the Solar Spectrum

Fraunhofer Designation	Wavelength (nm)	Element	Fraunhofer Designation	Wavelength (nm)	Element
A	759.4	O_2*	b_2	517.3	Mg
B	686.7	O_2*	b_3	516.9	Fe
C	656.3	H-alpha	F	486.1	H-beta
D_1	589.6	Na	G'	434.0	H-gamma
D_2	589.0	Na	G	430.8	Fe + CH
D_3	587.6	He	h	410.2	H-delta
E	527.0	Fe	H	396.8	Ca
b_1	518.4	Mg	K	393.4	Ca

Notes. *Abbreviations*: calcium (Ca), carbon–hydrogen molecule (CH), helium (He), hydrogen (H), iron (Fe), magnesium (Mg), oxygen (O), sodium (Na). An asterisk (*) indicates a terrestrial line, i.e., a spectral line produced by absorption of solar light by atoms or molecules in the Earth's atmosphere.

Worksheet, Activity 8: Spectral Lines and the Chemistry of the Sun

Name _____

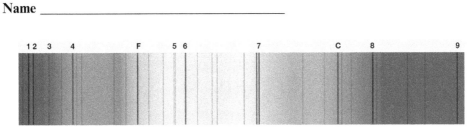

Figure 8.2. The solar spectrum, featuring the dark Fraunhofer lines. If in color, violet would appear at the left end of the spectrum, red at the right end; that is, wavelength increases from left to right.

1. (a) Separation Δd between C and F lines (to the nearest tenth of a mm) = _____ mm

   (b) Wavelength difference Δw between C and F lines = _____ nm

   (c) Image scale s = _____ nm mm^{-1}

2, 3.

Table 8.2. Spectral Line Data Table

Line number	Reference line (C or F)	Δmm	Δnm	w (nm)	Fraunhofer designation	Fraunhofer wavelength (nm)	Element
1							
2							
3							
4							
5							
6							
7							
8							
9							

Introduction to Stars and Planets
An activities-based exploration
Alan Hirshfeld

Activity 9

Is the Sun on Fire?

Preview

Before the 20th-century advances in our understanding of the atom, the source of the Sun's energy was a mystery. One suggestion was that the Sun is on fire, consuming some combustible substance. The test of this or any proposed energy-generation process is whether it can account for both the Sun's observed energy output and its several-billion-year lifetime.

9.1 The Energy Enigma

For centuries, astronomers wondered how the Sun generates its colossal energy outflow and speculated about its age and its longevity. By the late-1800s, geological evidence had accumulated that Earth is extremely old and that terrestrial life forms had evolved over many millions, if not billions, of years. Astronomers were hard-pressed to come up with a solar energy source that could have powered the Sun at its observed luminosity for such long durations of time. This, in an age when the structure of the atom was still a mystery and the concept of nuclear energy lay decades in the future. Nevertheless, various proposals were put forward for the origin of the Sun's radiance, including the accretion of high-speed meteors onto the solar surface, the slow contraction and heating of the solar interior, and chemical combustion of some highly volatile fuel.

Any theory of solar energy generation stands or falls on a pair of requirements:
- it must be capable of generating an amount of energy equivalent to the Sun's observed luminosity; and
- it must be able to do so, *without interruption*, for billions of years. (There is no evidence whatsoever of extended dips in solar-energy inflow to the Earth, that is, the Sun's luminosity has apparently been constant over the Earth's history.)

To put the various energy-generation theories to the test, we will make use of two solar attributes we had previously derived:

- the Sun's mass M_s, approximately 2×10^{30} kilograms (kg); and
- the Sun's luminosity L_s, about 3.86×10^{26} watts (W).

Spectral analysis reveals that the Sun contains a significant amount of nature's simplest chemical element: hydrogen. This lighter-than-air, highly flammable gas had been used to lift blimps and dirigibles until the infamous explosion of the giant airship *Hindenburg* in 1937. Might the combustion of hydrogen gas be the source of the Sun's light? Or to put it more plainly, *is the Sun on fire?*

9.2 Chemical Combustion

Combustion, or burning, is an energy-producing ("exothermic") chemical reaction involving the ignition of a fuel in the presence of oxygen. For example, given a sufficiently high temperature or an electrical spark, two hydrogen molecules (H_2) and one oxygen molecule (O_2) will interact to form a pair of water-vapor molecules (H_2O). Accompanying this reaction is a release of heat energy, specifically, 120 million (1.2×10^8) joules of energy per kilogram of hydrogen undergoing the reaction.

The joule (J) is a standard unit of energy and is equivalent to a watt-second; for instance, a 40 W light bulb shining for 10 s consumes $40 \times 10 = 400$ joules of energy from the electrical power grid. By reverse logic, the watt is equivalent to a joule per second; thus the Sun's luminosity, 3.86×10^{26} watts (W), can be written 3.86×10^{26} joules per second ($J\ s^{-1}$).

1. Since the hydrogen combustion reaction described above requires two hydrogen atoms for every oxygen atom, we will make the (erroneous!) assumption that there is sufficient oxygen present in the Sun to burn up all of the Sun's hydrogen. In other words, we claim that the Sun's material makeup reflects this 2:1 ratio of hydrogen to oxygen, and contains no other elements. Assuming this 2:1 ratio, of the Sun's overall mass M_s, how much is hydrogen and how much is oxygen?

2. Given that hydrogen combustion releases $1.2 \times 10^8\ J\ kg^{-1}$, how many kilograms of hydrogen would have to burn up *each second* to generate an energy equal to the Sun's observed luminosity L_s?

3. Now we estimate the lifetime of the hydrogen-combustion process, that is, how long can it run before exhausting *all* of the Sun's hydrogen fuel? The first step is to compute the Sun's overall energy-production capacity E from this process. To do this, multiply the total mass of hydrogen in the Sun from part 1 (in kg) by the amount of energy produced in the combustion reaction (in $J\ kg^{-1}$). Your answer will come out in units of joules.

4. The second step in assessing the lifetime of solar hydrogen combustion is to divide the Sun's overall energy-production capacity E from part 3 by its luminosity L_s, the rate at which the Sun consumes—radiates away—its

overall store of energy. Convert your answer from units of seconds into years, given that 1 yr $= 3.15 \times 10^7$ s.

5. How does your estimated solar lifetime from hydrogen combustion in part 4 compare to the Sun's purported lifetime of billions of years? Can hydrogen combustion account for the Sun's energy output? Explain.

9.3 Conclusion

In truth, we grossly overestimated the amount of oxygen in the Sun; there is barely enough to burn hydrogen for several decades. A similar step-by-step analysis can be applied to the combustion of any other substance: wood, gasoline, coal, even bananas! In 1937, solar astronomer Donald Menzel picked up on a grammatical error in the *Chicago Daily News*, which, if read literally, implied that the Sun derived its light by consuming bananas. Knowing a banana's approximate calorie count, Menzel determined that the Sun would have to combust the fruity-equivalent of the Earth's entire mass *every second* to match the observed solar luminosity. He mischievously—and prophetically—urged astronomers to instead "theorize about sub-atomic chemical reactions that involve the hearts of atoms."

Comparable tests have been applied to other historical candidates for solar energy production, such as meteor accretion, gravitational contraction, and nuclear fission. All fail for the identical reason: they cannot power the Sun continually for the billions of years we know the Sun has been shining. Only after mid-20th-century scientists probed the workings of the atomic nucleus did they find the source of the Sun's long-lasting brilliance.

Worksheet, Activity 9: Is the Sun on Fire?

Name _____

1. Mass of Sun's hydrogen = _____ kg

   Mass of Sun's oxygen = _____ kg

2. Hydrogen consumed each second = _____ kg

3. E = _____ J

4. E/L_s = _____ s = _____ yr

5. _____
   _____
   _____
   _____

Introduction to Stars and Planets
An activities-based exploration
Alan Hirshfeld

Activity 10

How Long Will the Sun Shine?

Preview

The Sun's overall life span is estimated from its mass and luminosity, assuming its radiant energy comes from the thermonuclear fusion of hydrogen into helium in the solar core.

10.1 Thermonuclear Fusion

Without the Sun's life-sustaining energy, Earth would be a dark, desolate world. Instead, it teems with a multitude of species, including our own. Once biologists, paleontologists, and geologists concluded that life on Earth had evolved over the span of several billion years, astronomers were hard-pressed to explain the longevity and the constancy of the Sun. What is the source of this immense stellar energy? Among the various mechanisms that had been put forward, such as meteor accretion, chemical combustion, or gravitational contraction, none could power the Sun continually for billions of years. The wellspring of solar energy, scientists speculated, must lie in some inherent quality of the particles in the ultra-hot depths of the Sun.

During the early decades of the 20th century, physicists applied the Sun's measured physical properties—mass, radius, luminosity, surface temperature, and chemical composition—to the latest theories governing the behavior of matter and energy. They concluded that the deep solar interior is sufficiently hot that fast-moving hydrogen nuclei might overcome their natural electrical repulsion and join together. Through a sequence of such mergers, these hydrogen nuclei would transform into nuclei of a different element, helium, in a process physicists call thermonuclear fusion.

Each time a helium nucleus is synthesized from hydrogen nuclei within the Sun's core, a tiny amount of matter is converted into energy (mostly gamma-rays), which, after a long and meandering rise through the solar interior, emerges into space as sunlight. Using a few fundamental physics concepts, we can estimate how many

doi:10.1088/2514-3433/abc249ch10
© IOP Publishing Ltd 2020

years the Sun can shine from this fusion-energy mechanism before it runs out of its core supply of hydrogen.

At the outset, it's important to recognize that chemical combustion of hydrogen and thermonuclear fusion of hydrogen are wholly different phenomena. Combustion rearranges *atoms* of one chemical compound into another; any heat energy produced is derived from the dissolution and subsequent re-formation of atomic or molecular bonds. Fusion, on the other hand, transforms atomic *nuclei* from one element into another, and releases far more energy per reaction than combustion does.

In the seething stellar hothouse where fusion occurs, electrons in an atom are shaken free of their host nuclei and move unfettered within a gaseous sea of particles. Thus, thermonuclear fusion is a strictly *nuclear* interaction, not an atomic one. So don't be misled if you come across the term "hydrogen burning," which is an astronomer's colloquialism for the fusion process; no chemical combustion occurs in fusion.

10.2 Input Data

The basic information we need to estimate the Sun's fusion-driven life span are:
- the makeup of a hydrogen (H) nucleus: **1 proton**.
- the makeup of a helium (He) nucleus: **2 protons plus 2 neutrons**.
- the mass of a hydrogen nucleus, which consists of a single subatomic particle called a proton: **1.0079 g**. In fact, this is *not* the mass of an individual nucleus, but the mass of a specified standard number of nuclei, and is termed an element's *atomic mass*. (A gram is one one-thousandth—1/1000—of a kilogram.)
- the mass of a helium nucleus: **4.0026 g**. As explained above, this number is actually helium's atomic mass.
- the Sun's mass: $\mathbf{2 \times 10^{33}}$ **g**. We computed the Sun's mass in a previous activity, only in kilograms instead of grams.
- the Sun's luminosity L, that is, the overall rate at which it releases energy: $\mathbf{4 \times 10^{33}}$ **ergs s^{-1}**. An erg is the unit of energy traditionally used by astronomers; we previously derived the Sun's luminosity, only in a different unit: watts.

10.3 Mass into Energy

It's straightforward to estimate the Sun's overall life span, assuming that its energy is derived from the thermonuclear fusion of the element hydrogen (H) into helium (He). To make 1 helium nucleus requires the fusion of 4 hydrogen nuclei, or protons. Two of these protons morph into neutrons by shedding their positive charge in the form of a subatomic particle called a positron and releasing an electrically-neutral fleck of matter called a neutrino.

From the list of basic data above, we note that the mass of 1 helium nucleus is *slightly less* than the combined mass of the 4 hydrogen nuclei that go into its creation. This tiny reduction in mass between the reaction material (hydrogen) and the final product (helium) represents matter that has been converted into energy during the fusion process. Our claim is that this jot of energy, multiplied by the immense number of hydrogen-to-helium fusions that occur every second in the Sun's hot core, tallies up to the Sun's observed luminosity.

1. Using the atomic mass data provided, compute the combined atomic mass of 4 hydrogen nuclei, in grams.
2. Subtract the atomic mass of 1 helium nucleus from the combined atomic mass of the 4 hydrogen nuclei in part 1.
3. Divide the mass difference from part 2 by the combined atomic mass of the 4 hydrogen nuclei from part 1. Enter this value, which we'll designate f, on the worksheet.

Your answer to part 3 represents the fraction f of the starting hydrogen mass that has been converted into energy while being fused into helium. This fraction f applies not only to the mass of 4 hydrogen nuclei undergoing fusion, but to *any* mass of hydrogen undergoing fusion. If, say, 1 g of hydrogen is fused into helium, the same fraction f of that 1 g will be converted into energy. Every time 1 g of hydrogen fuses into helium in the Sun's core, the Sun's mass diminishes by a fraction f of a gram; what had once been material substance is being radiated away in the form of luminous energy!

We next use Einstein's iconic equation, $E = mc^2$, to compute how much energy is produced each time a *gram* of hydrogen undergoes fusion. In this case, the mass m in Einstein's equation is equal to the quantity f, in grams, from part 3. If the constant c, the speed of light, is expressed as 3×10^{10} centimeters per second (cm s^{-1}), your answer E will come out in the chosen unit of energy for this activity: ergs. (In a previous activity, we had used a related energy unit called the joule.)

4. Using Einstein's equation, $E = mc^2$, calculate how much energy is produced when 1 g of hydrogen is fused into helium.

Now let's compute how much hydrogen is being consumed by the Sun's fusion process every second. If the Sun emits energy at the rate of L ergs per second, and…

… if each fused gram of hydrogen contributes to this radiant emission an amount of energy E ergs…

… then the number of grams of hydrogen being fused every second inside the Sun is simply L divided by E.

5. (a) Compute L/E. (b) Of this mass of fused hydrogen, a fraction f is consumed—turned into energy—every second. Multiply your answer to part 5(a) by the fraction f from part 3. Express your answer first in grams, then convert that number to metric tons. (1 metric ton = 1 million (10^6) g; a heavy freight train weighs in at around 10,000 metric tons.) Your answer to this part represents the enormous amount of mass the Sun loses *every second* as the result of fusion-based energy production!

10.4 How Long Can Fusion Last?

To assess whether fusion could indeed power the Sun for the requisite 4.6 billion years the Sun is presumed to have been shining, we now compute the total amount of energy the Sun, in theory, can generate by fusion before it exhausts its hydrogen supply. For simplicity, let's assume that the Sun is made completely of hydrogen, that is, we'll ignore the much lesser amount of helium presently in the Sun.

In figuring out how much solar hydrogen is ultimately available for fusion, we have to recognize that fusion takes place only in the Sun's core, where the temperature is sufficiently high for the fusion mechanism to proceed. Based on current estimates, about one-tenth (0.1) of the Sun's mass lies within its energy-generating core. We will assume that this "core mass" represents how many grams of hydrogen are available to undergo fusion during the Sun's lifetime.

6. Apply the fraction one-tenth to the Sun's mass (from the data list above) to determine the Sun's core mass, that is, its overall "fuel supply."

7. While *all* of the solar core mass in part 6 will undergo fusion, only the fraction f of that mass will be converted into energy. (The rest remains in the form of matter, specifically, helium nuclei, positrons, and neutrinos.) Apply the fraction f to your answer in part 6 to compute the total mass M of hydrogen that will be transformed into energy during the Sun's lifetime.

8. Again, we turn to Einstein's equation, $E_M = Mc^2$, this time to compute the total supply of energy E_M that can be generated by the available mass M of hydrogen found in part 7. (Here, we adopt the symbol E_M to distinguish this energy from one we dealt with earlier.)

10.5 The Life Span of the Sun

We are now ready to calculate how long the Sun's fusion energy supply E_M will last. The life span of the Sun is found by dividing (i) the overall energy content E_M of the hydrogen fuel in its core by (ii) the rate at which it radiates away this energy, that is, its luminosity L.

9. Using your values of E_M from part 8 and the Sun's luminosity L from the data list above, compute E_M/L, the potential life span of the Sun. Your answer comes out in the unit of seconds.

10. (a) Given that there are 3.15×10^7 s in a year, convert the Sun's computed life span in part 9 from seconds into years. (b) Write out your final numerical answer in words, that is, use appropriate terms like million, billion, etc.

11. Does your answer to part 10 support the claim that the solar fusion process *can* last for at least the 4.6 billion years we believe the Sun has been shining? If not, check your calculations for errors.

Worksheet, Activity 10: How Long Will the Sun Shine?

Name _____

1. Mass (4H) = _____ g

2. Mass (4H−1He) = _____ g

3. Fraction of hydrogen mass converted into energy (f) = _____

4. Fusion energy from 1 g of hydrogen = _____ ergs

5. (a) Hydrogen fused per second (L/E) = _____ g

   (b) Hydrogen converted to energy ("consumed") = _____ g

       = _____ metric tons

6. Mass of Sun's core = _____ g

7. Mass of Sun's core convertible to energy (M) = _____ g

8. Sun's energy supply, past and future (E_M) = _____ ergs

9. Potential life span of the Sun, in seconds (E_M/L) = _____ s

10. (a) Life span of the Sun, in years = _____ yr

    (b) _____

11. _____

Part II

The Stars

Introduction to Stars and Planets
An activities-based exploration
Alan Hirshfeld

Activity 11

The Distances of Stars: Stellar Parallax

Preview

The distances of stars are determined by measuring their annual parallax, the cyclical shift in a star's sky position when viewed from opposite extremes of the Earth's wide-swinging orbit. The vastness of the starry realm is revealed, as is the illusory clustering of stars within constellations.

11.1 Stellar Parallax

Ancient Greek astronomers situated the Earth at the center of the visible universe, as they perceived it more than two thousand years ago. This geocentric model of the cosmos held sway among scholars until the Polish astronomer Nicolaus Copernicus proposed his Sun-centered (heliocentric) alternative in the mid-1500s. Copernicus acknowledged one problematic feature of his revolutionary rearrangement of the heavens: if the Earth circles the Sun, as he claimed, then the stars would oscillate slightly in the night sky over the course of the year, a subtle reflection of our planet's changing location. Over a six month interval the Earth stands at diametrically opposite places in its orbit, and we view the stars from these widely separated points in space, as depicted in Figure 11.1. Thus, the sky coordinates of a star tonight will differ from the star's coordinates half a year hence. This is the well-known *parallax* phenomenon: the apparent shift in an object's position when viewed from different vantage points.

It's easy to demonstrate the parallax phenomenon: hold up your finger several inches in front of your face, close your left eye and view your finger with only your right eye, then close your right eye and view it only with your left eye. Repeat the process several times. The apparent side-to-side jump of your finger against the fixed backdrop is the parallax effect. Carry out the observation again, only with your finger held at arm's length; the parallax shift this time should be noticeably smaller. In practice, an object's parallax shift can be measured and its corresponding distance computed: the smaller the parallax, the farther the object. Indeed, as Copernicus

Figure 11.1. Stellar parallax is the apparent shift in a star's sky position when viewed from opposite ends of the Earth's orbit, here, in June and December. The star's parallax shift over six months is represented by the angle between the dashed lines. The figure is not to scale; even "nearby" stars are extremely far away. (Credit: NASA, ESA, A. Feilds/STSci. CC BY 4.0.)

knew, measurement of a star's parallax over half a year would provide a direct geometric means to gauge the star's distance.

Copernicus was well aware that stellar parallax had never been observed, a circumstance that lent credibility to the concept of a central, stationary Earth, as in the reigning geocentric cosmic model. He countered this seeming flaw in his theory by suggesting that the stars are so remote that their parallax oscillations are too small to be seen by eye. With the advent of the telescope in the 1600s, astronomers took up the search for these elusive star wobbles. But it would not be until 1838 that German astronomer-mathematician Friedrich Bessel succeeded in measuring the first stellar parallax. As Copernicus had conjectured, the parallax shift was exceedingly small, confirming what astronomers had already suspected: the distances of stars are truly vast compared to the size of our solar system. Today, the parallaxes—and corresponding distances—of more than a billion stars have been measured. This activity will acquaint you with the basics of distance estimation using parallax measurements.

11.2 Parallax Simulation

Figure 11.2 shows a series of simulated "photographs" of the same field of stars. Photo 1 depicts the field of stars as it appeared one night in the sky; Photo 2 shows the same field six months later; and Photo 3 displays the stars 6 months after Photo 2 (one year after the original photograph, Photo 1).

Although the stars in Figure 11.2 might look alike, one of these stars is actually much closer to the Earth than the others. You can determine which star is the closer one by recalling how stellar parallax relates to distance, this time in the reverse sense: the larger a star's parallax, the closer the star.

Parallax Simulation

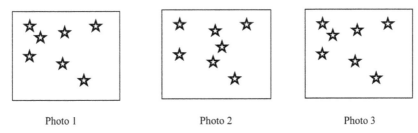

| Photo 1 | Photo 2 | Photo 3 |

Figure 11.2. Simulated images of a field of stars taken at three different points in time.

1. How does the Earth's position around the Sun differ from Photo 1 to Photo 2 to Photo 3?
2. How can you tell which star in the photographs is closest to the Earth?

11.3 Bessel's Star

In 1838 October, Friedrich Bessel announced the results of his year-long quest to measure the parallax of his target star in the constellation of Cygnus. The result confirmed what astronomers had by then suspected: even the nearest stars to our solar system are exceedingly distant. For example, the parallax angle of Bessel's star (officially, 61 Cygni) is a mere 0.00008 degrees, placing the star more than 700,000 au away!

To bring such enormous stellar distances "down to Earth," let's reduce outer space to a more comprehensible size. Imagine shrinking the cosmos uniformly such that the radius of the Earth's orbit, 1 au, matches the separation between your eyes, typically around 6 cm. That is, the Sun would coincide with one eye and the Earth would coincide with the other.

3. (a) On the shrunken scale described above, how many centimeters away from your face would Bessel's star lie? (b) Covert your answer to the previous part from centimeters into meters and then into kilometers. (There are 100 cm in a meter and 1000 m in a kilometer.)
4. Given your answer to part 3, explain how the representation of the Earth's orbit and the "nearby" star in Figure 11.1 would change if drawn to the actual scale of stellar distances.

11.4 The 3D Cosmos

Astronomers have divided the night sky into 88 constellations: arrangements of stars that have been associated with mythological characters, animals, or sometimes everyday objects. While a constellation's stars appear close to one another in the sky, they are often widely separated from each other in space. Once astronomers succeeded in measuring stellar parallaxes, the cosmic third dimension—distance—was added to our age-old, two-dimensional perspective on the night sky. In mapping out the stars near our solar system, astronomers adopted a distance unit called the *parsec*, which is defined as the reciprocal of a parallax angle of 1 arcsecond, or

Figure 11.3. The constellation of Orion. The reddish star Betelgeuse (upper left) and the bluish-white star Rigel (lower right) are among the brightest stars in the night sky. Below Orion's "belt"—the diagonal triplet of stars at center—is the Orion Nebula, a glowing interstellar gas cloud. (Credit: NASA/JPL-Caltech/IRAS/H. McCallon.)

1/3600 of a degree. Thus a star whose parallax angle is 1 arcsec lies 1 parsec from the Earth. Or in general, a star's distance d, in parsecs, can be computed by taking the reciprocal of its parallax angle p, in arcseconds:

$$d = \frac{1}{p}. \tag{11.1}$$

5. Figure 11.3 is a photograph of the constellation Orion, which is prominent in the Northern Hemisphere's winter night sky. Table 11.1 on the worksheet lists the parallaxes (in arcseconds) of Orion's seven brightest stars. Use Equation (11.1) to compute the distance to each of these stars. Write your answers in the table.

Figure 11.4 on the worksheet shows two views of Orion's major stars. On the left appears the two-dimensional view that we see in the night sky. To its right, we will plot a space-based "side view" of these same stars, which will show how they would appear from some imaginary outlook hundreds of parsecs from Earth. This latter view will reveal whether Orion's stars, which appear close to one another in the sky, are, in fact, close to one another in space.

6. For each star in Table 11.1, plot a data point in Figure 11.4 along that star's horizontal line at the star's listed distance. After you've plotted all seven stars, fold back the worksheet page along the indicated line so that the two parts of the page are perpendicular to each other. Here is a three-dimensional rendering of Orion's stars!
7. Which of Orion's stars are grouped together in space? Which are not?

Worksheet, Activity 11: The Distances of Stars: Stellar Parallax

Name _____

1. _____
   _____
2. _____
   _____
3. (a) _____ cm
    (b) _____ m = _____ km
4. _____
   _____
5.
Table 11.1. Parallaxes of Orion Stars

Star	Parallax (arcsec)	Distance (parsec)
Betelgeuse	0.0066	
Bellatrix	0.0129	
Mintaka	0.0047	
Alnilam	0.0017	
Alnitak	0.0044	
Rigel	0.0038	
Saiph	0.0050	

6.

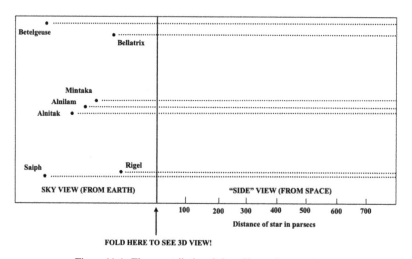

Figure 11.4. The constellation Orion: Sky and space views.

7. _____
   _____

Introduction to Stars and Planets
An activities-based exploration
Alan Hirshfeld

Activity 12

Weighing a Star: Binary Stars and Stellar Mass

"The visibility of countless stars is no argument against the invisibility of countless others."—Astronomer Friedrich Bessel, in an 1846 letter, after concluding that the star Sirius is paired with an unseen companion star.

"Prof. Bond [at Harvard] communicates the discovery of a Companion of Sirius, made on the evening of Jan. 31 by Mr. Clark, with his new object-glass of 18½ inches aperture."—*Monthly Notices of the Royal Astronomical Society*, 1862 March 14.

Preview

Kepler's third law, relating the orbital period of a celestial body and the radius of its orbit, is used to determine the mass of individual stars in binary systems. The method is applied to the bright star Sirius and its faint companion, a dense, Earth-sized star known as a white dwarf.

12.1 Binary Stars

In 1803, the eminent English astronomer William Herschel concluded that at least some double stars—pairs of stars close together in the sky—are, in fact, binary systems: stars in orbit around one another. Astronomers realized that observations of binary stars would allow them to determine a stellar property that was otherwise measurable only for the Sun: a star's mass. The basis of stellar mass measurement dates to the mid-1600s, after Isaac Newton derived Kepler's mathematical laws of orbital motion from his own Universal Law of Gravitation. The upshot is that any orbiting system of objects bound together by gravity will conform to Kepler's laws, whether that system is the Moon orbiting the Earth, a planet orbiting the Sun, or a binary star whose members orbit each other. Therefore, careful measurements of the

doi:10.1088/2514-3433/abc249ch12   12-1   © IOP Publishing Ltd 2020

orbital radius and orbital period of a binary star could lead to a determination of stellar mass via Kepler's third law, which interrelates these three quantities.

Some four decades after Herschel's work, German astronomer Friedrich Bessel found that the bright star Sirius displays a wobbling movement in the sky over many years, as though it were dragging along a massive, but invisible, companion star. That furtive body, all but lost in Sirius's glare, was discovered accidentally in 1862 by telescope maker Alvan Graham Clark, in Cambridge, Massachusetts, while testing a new telescope lens. It was a star of exquisite strangeness, unlike any that had ever been seen. It was massive enough to tug heavy Sirius from its regular, straight-line path, yet barely luminous enough to be seen through a powerful telescope: a *white dwarf* star.

In this activity, you will "observe" the Sirius star system and use Kepler's third law to determine the mass of Sirius itself, as well as that of its elusive companion, known as Sirius B. Kepler's Third Law is typically written in the familiar short form $P^2 = a^3$ for objects orbiting the Sun. A more general form, applicable to celestial bodies outside the solar system, is $M \times P^2 = a^3$ or, with a slight algebraic manipulation:

$$M = \frac{a^3}{P^2}. \tag{12.1}$$

Here M stands for the *combined* mass of the orbiting star-pair—Sirius plus its companion—expressed as a multiple of the Sun's mass, a unit called a *solar mass*. The period P is expressed in years and the orbital radius a (technically, the *semimajor axis*) is expressed in astronomical units (au). In the steps below, you will first determine the orbital period of the Sirius system, then the orbital radius, and finally the combined and individual masses of the stars.

12.2 Orbital Period

Figure 12.1 shows the orbital path of Sirius B around Sirius itself. This orbit is based on decades of telescopic study, accumulating a sufficient number of observations to define the gradual movement of Sirius B. In reality, the two stars orbit each other,

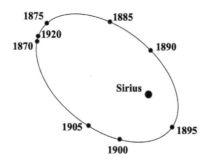

Figure 12.1. Orbital path of Sirius B (small dots, each with the year of observation) around the star Sirius (large dot).

but for clarity the orbit is displayed as though Sirius is fixed in place and Sirius B moves around it.

1. (a) Study the labeled years of observation and use them to estimate the orbital period of Sirius B, in years. (b) Explain how you arrived at your answer.

12.3 Orbital Radius and Combined Mass of Sirius and Sirius B

The orbital radius of the Sirius system, *as seen from the Earth*, spans an angle of about 7.6 arcsec (0.002°). However, Kepler's third law requires, not the angular measure of the orbit, but the *actual* radius of the orbit. The sector formula that you used in previous activities can be used here as well to make the needed conversion:

$$a = \frac{d \times \theta}{57.3},$$
(12.2)

where a represents the actual orbital radius, in au; θ represents the angular span of the orbital radius, in degrees; and d represents Sirius's distance, in au. The interrelationship among these quantities is shown in Figure 12.2.

The distance d in Equation (12.2) can be computed from Sirius's parallax, its change in position as seen from opposite ends of Earth's orbit: the closer the star, the larger its parallax. Sirius is relatively nearby, as stars go. Therefore, it's no surprise that its parallax was measured as early as the 1840s and is now known quite accurately: 0.379 arcsec.

As seen in a previous activity, the relationship between a star's distance d and its parallax p is given by the simple formula $d = 1/p$, where the distance d is measured in parsecs (1 parsec = 3.26 lt-yr) and the parallax p is measured in arcseconds.

2. Use the parallax formula in the previous paragraph to compute the distance of Sirius, in parsecs. (Don't confuse the period symbol P in Kepler's third law with the parallax symbol p.)
3. Convert the distance of Sirius from parsecs into au, given that 1 parsec = 206,265 au.
4. Use Equation (12.2) to compute the orbital radius a of the Sirius system, in au.
5. Substitute your computed values for the period P and the orbital radius a into Kepler's third law, Equation (12.1), and solve for the combined mass M of Sirius and its companion Sirius B, in solar masses.

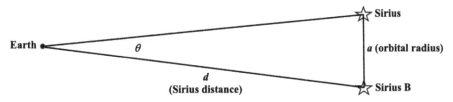

Figure 12.2. "Sector view" of the angular span θ, orbital radius a, and distance d of the Sirius binary system relative to Earth. (Not to scale.)

12-3

12.4 Masses of the Individual Stars

Using some elementary physics, it's possible to take the combined mass of the Sirius binary system and separate out the individual masses of Sirius and its companion, Sirius B. Figure 12.3 depicts the movement of the Sirius pair through space over many decades. The large dots indicate Sirius's position and the smaller dots the position of Sirius B. Corresponding dots for each year of observation are connected by a line segment. Tracing the figure from top to bottom, notice that as the orbiting system moves through space, the two stars oscillate around an imaginary straight line between them. This line defines the system's *center of mass*.

Here's where the physics comes in. The heavier star (Sirius) always lies closer to the center-of-mass line than the lighter star (Sirius B). And the ratio of Sirius's mass to that of its companion, M_s/M_c, will be equal to the ratio of the companion's distance from the center-of-mass line to that of Sirius, r_c/r_s. That is:

$$\frac{M_s}{M_c} = \frac{r_c}{r_s}. \tag{12.3}$$

The same principle explains why, in order to balance a seesaw, the heavier person must sit closer to the central pivot than does the lighter person.

6. To begin to solve Equation (12.3), we first have to measure r_c and r_s. In Figure 12.3, choose any one of the line segments that connect the stars for a given year of observation. Align a slip of paper alongside the segment. With your pencil, mark off the length of just that portion of the segment that extends from the center-of-mass line to Sirius. This is the "Sirius segment" r_s. Now slide your marked paper over to the portion of the segment that connects the center-of-mass line to the companion star, Sirius B. Approximately how many of these marked "Sirius segments" does it take to trace out the corresponding "companion segment?" The answer to this question is precisely the ratio we're looking for: r_c/r_s. And according to Equation (12.3), this ratio is equal to the mass ratio, M_s/M_c. You should carry out this procedure for several line segments in Figure 12.3 to confirm that your answers for r_c / r_s are consistent.

7. In parts 5 and 6, we determined the sum of the masses of Sirius and its companion, as well as the ratio of the masses of the individual stars. Now, using either simple algebra or trial and error, you can compute the individual masses of Sirius and Sirius B, in solar masses. For example, suppose $a + b = 12$ and $a / b = 3$. What are the values of a and b? Solution: $a / b = 3$ becomes $a = 3b$; substitute into the first equation $a + b = 3b + b = 12$, so $b = 3$, and therefore $a = 9$.

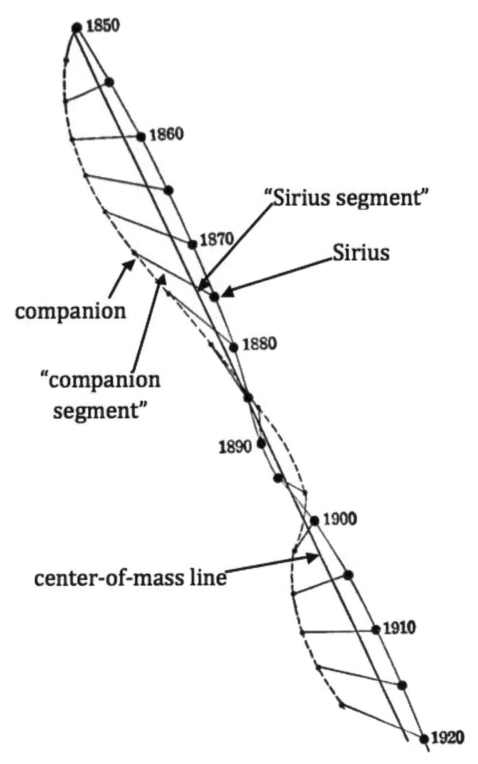

Figure 12.3. The oscillating paths of Sirius and its companion, Sirius B, through space.

Worksheet, Activity 12: Weighing a Star: Binary Stars and Stellar Mass

Name _____

1. (a) Orbital period = _____ yr

    (b) _____
    _____
    _____

2. Distance of Sirius d = _____ parsec

3. Distance of Sirius d = _____ au

4. Orbital radius a = _____ au

5. Combined mass M = _____ solar masses

6. Companion segment = _____ Sirius segments

7. Mass of Sirius = _____ solar masses

    Mass of Sirius B = _____ solar masses

Introduction to Stars and Planets
An activities-based exploration
Alan Hirshfeld

Activity 13

The Hertzsprung–Russell Diagram

Preview

The Hertzsprung–Russell, or HR, diagram displays the relationship between two fundamental physical parameters of stars—luminosity and surface temperature—and is one of the most useful graphical instruments in the astronomer's tool kit. An HR diagram constructed from data collected for the visually brightest stars and the nearest stars reveals important distinctions between these categories and highlights the observational pitfall known as selection effects.

13.1 Introduction

One way to begin a scientific analysis of a class of objects is to look for commonalities in their various physical properties—size, mass, density, etc.—and then develop an appropriate classification scheme. In addition, we can graph observed data for a representative sample of objects, then inspect the resulting display for correlations among these measures. For example, in a large, randomly selected sample of people, we expect to find a relationship between, say, people's height and their shoe size: generally, tall people tend to have larger feet. In astronomy, there is a wide variety of such analytic techniques to investigate the inherent properties of stars, planets, and other celestial bodies.

13.2 Stellar Magnitudes and Spectral Types

Astronomers have long applied quantification, classification, and correlation to the menagerie of cosmic forms that inhabit our universe. For instance, by the early 20th century, astronomers had reliably determined the distances of several dozen stars by the parallax method, whereby a slight annual fluctuation in a star's position reveals its remoteness from the Sun. Astronomers next applied the inverse square law, which quantifies the decrease in a star's light intensity with distance, to deduce each star's luminosity, or overall energy output.

doi:10.1088/2514-3433/abc249ch13

13-1

© IOP Publishing Ltd 2020

Traditionally, stellar brightness is expressed as a number called the *magnitude*. A star's *apparent magnitude* **m** denotes how bright the star appears from the Earth, whereas its *absolute magnitude* **M** expresses how bright the star would appear if relocated to a distance of 10 parsecs, or about 32.6 lt-yr. (A parsec is the distance of a star whose parallax angle is one arcsecond.) The astronomical magnitude system is "backwards": the smaller the magnitude number, the brighter the star. Thus, a star of magnitude **2** outshines a star of magnitude **5**, but is dimmer than a star of magnitude **−1.5**.

Each one-magnitude *difference* between stars represents a brightness *ratio* of about 2.5. Thus, a star of magnitude **4** is 2.5 times brighter than one of magnitude **5**, while a star of magnitude **3** is $2.5 \times 2.5 \times 2.5$, or about 16, times brighter than one of magnitude **6**. A 5 magnitude difference between stars yields a brightness ratio of $2.5 \times 2.5 \times 2.5 \times 2.5 \times 2.5$, that is, $(2.5)^5$, or about 100.

In this activity, the absolute magnitude **M** stands in for stellar luminosity: both are indicators of how much energy a star emits relative to other stars. For example, the Sun's apparent magnitude is **−26.7**, overwhelming the feeble rays of every other celestial object during the daytime; yet, were it shunted off to the standardized distance of 10 parsecs, the Sun's absolute magnitude would register at **4.8**, an average light-emitter among the general stellar population.

Also around the dawn of the 20th century, researchers at Harvard University began to classify stars according to the appearance of their spectrum, adopting alphabetic designations—A, B, and so on—that have since become standard. They soon realized that the major observed variations among these spectral types stemmed, not so much from differences in stars' chemical composition, but from differences in their *surface temperature*. As a result, the Harvard astronomers rearranged their alphabetical spectral types into a tabulation by stellar temperature, from hottest to coolest: O, B, A, F, G, K, and M. Each spectral type was subsequently divided by temperature, into ten parts, indicated by a digit from 0 to 9 (for example, A0, A1, A2, ..., A8, A9, then F0, F1, F2, and so on). The Sun is spectral type G2.

Once astronomers gained a handle on this pair of stellar characteristics—luminosity and surface temperature, or their proxies, absolute magnitude and spectral type—they wondered whether there is any interrelationship between the two. Might the most energetic stars, say, be associated with only certain spectral types? Do all G2-type stars have the same luminosity as the Sun? Working independently, Danish astronomer Ejnar Hertzsprung and his American counterpart, Henry Norris Russell, explored the issue by measuring and graphing these fundamental stellar properties for a sample of stars. The resultant graph, called the Hertzsprung–Russell, or HR, diagram, has become a critical tool in the scientific study of stars and their evolution over time. In this activity, you will follow the path blazed by Hertzsprung and Russell, but using modern measurements.

13.3 Plotting and Reading the HR Diagram

1. Table 13.1 lists physical data for two categories of stars: the visually brightest stars in the sky; and the nearest stars, those within about 10 lt-yr of the solar

Table 13.1. Data for the Visually Brightest and the Nearest Stars: Star Name; Distance in Light-years; Apparent Magnitude *m*; Absolute Magnitude *M*; and Spectral Type

	Brightest Stars (●)				Nearest Stars (○)				
Name	Dist (lt-yr)	App mag m	Abs mag M	Spec type	Name	Dist (lt-yr)	App mag m	Abs mag M	Spec Type
Sun	—	−26.7	4.8	G2 V	Sun	—	−26.7	4.8	G2 V
Sirius A	8.6	−1.46	1.4	A1 V	Proxima Centauri	4.2	11.05	15.5	M6 V
Canopus	74	−0.72	−2.5	A9 II	Alpha Centauri A	4.3	−0.01	4.4	G2 V
Alpha Centauri A	4.3	−0.01	4.4	G2 V	Alpha Centauri B	4.3	1.33	5.7	K0 V
Alpha Centauri B	4.3	1.33	5.7	K0 V	Barnard's Star	6.0	9.54	13.2	M4 V
Arcturus	34	−0.04	0.2	K2 III	Wolf 359	7.7	13.53	16.7	M6 V
Vega	25	0.03	0.6	A0 V	Lalande 21185	8.2	7.50	10.5	M2 V
Capella	41	0.08	0.4	G6 III	UV Ceti A	8.4	12.52	15.5	M6 V
Rigel	~1400	0.12	−8.1	B8 Ia	UV Ceti B	8.4	13.02	16.0	M6 V
Procyon A	11.4	0.38	2.6	F5 IV–V	Sirius A	8.6	−1.46	1.4	A1 V
Achernar	69	0.46	−1.3	B3 V	Sirius B	8.6	8.3	11.2	D(A2)
Betelgeuse	~1400	0.50	−7.2	M2 Iab	Ross 154	9.4	10.45	13.1	M4 V
Hadar	320	0.61	−4.4	B1 III	Ross 248	10.4	12.29	14.8	M5 V
Acrux	510	0.76	−4.6	B1 IV	Epsilon Eridani	10.8	3.73	6.1	K2 V
Altair	16	0.77	2.3	A7 V	Lacaille 9352	10.8	7.34	9.8	M2 V
Aldebaran	60	0.85	−0.3	K5 III	Ross 128	10.9	11.10	13.5	M4 V
Antares	~520	0.96	−5.2	M2 Iab	61 Cygni A	11.1	5.2	7.6	K4 V
Spica	220	0.98	−3.2	B1 V	61 Cygni B	11.1	6.03	8.4	K5 V
Pollux	40	1.14	0.7	K0 III	Epsilon Indi	11.2	4.68	7.0	K3 V
Fomalhaut	22	1.16	2.0	A3 V	BD +43 44 A	11.2	8.08	10.4	M1 V
Becrux	460	1.25	−4.7	B1 III	BD +43 44 B	11.2	11.06	13.4	M4 V
Deneb	1500	1.25	−7.2	A2 Ia	Procyon A	11.4	0.38	2.6	F5 IV–V
Regulus	69	1.35	−0.3	B7 V	Procyon B	11.4	10.7	13.0	D(A)
Adhara	570	1.50	−4.8	B2 II	BD +59 1915 A	11.6	8.90	11.2	M3 V
Castor	49	1.57	0.5	A1 V	BD +59 1915 B	11.6	9.69	11.9	M4 V
Gacrux	120	1.63	−1.2	M4 III	CoD −36 15693	11.7	7.35	9.6	M1 V

Notes. Roman numerals in the spectral type indicate the size of the star— Ia and Iab, supergiant stars; II and III, giants; IV, subgiants; and V, main sequence stars— while the letter D before the spectral type signifies a white dwarf star. A tilde symbol (~) means that the listed value is approximate.

system. For each of the two data sets, plot the absolute magnitude versus the spectral type on the axes of Figure 13.1 on the worksheet. Use a filled circle (●) to represent the visually brightest stars and an open circle (o) to represent the nearest stars. A letter "D" before the regular spectral-type letter indicates a collapsed star known as a *white dwarf*. And remember, the *smaller* the magnitude number, the *brighter* the star!

2. Observe that the majority of the nearest and (to a lesser extent) the visually brightest stars fall along a diagonal in the HR diagram, a band that astronomers have named the *main sequence*. Describe the general luminosity and temperature profile of stars in the (a) upper part of the main sequence (b) lower part of the main sequence.

3. Another group of stars falls in the upper right portion of the HR diagram. These are *red giant* and *red supergiant* stars. Draw a circle around this group of stars in Figure 13.1 on the worksheet. (a) Describe the general luminosity and temperature profile of red giants and supergiants. (b) Can you explain why astronomers believe these stars to be very large? Hint: A "cool" star— one whose surface temperature is, say, 3500°C—radiates less energy per square meter of its surface than does a hot star.

4. In the lower left portion of the HR diagram are white dwarf stars. Circle these in Figure 12.1. (a) Describe the general luminosity and temperature profile of white dwarfs. (b) Can you explain why astronomers believe these stars to be very small, around the size of the Earth? Hint: A hot star radiates more energy per square meter of its surface than does a cool star.

5. (a) How does the overall distribution of the visually brightest stars (the "●" symbols) differ from that of the nearest stars (the "o" symbols) in the HR diagram? (b) What does this imply about the fundamental properties of each of these two categories of stars?

6. Which star in Table 13.1 appears brightest in the *night* sky? (It's not the Sun!)

7. (a) Which star in Table 13.1 is farthest from the solar system? (b) Why does such a faraway star nevertheless appear so bright in the night sky?

13.4 Selection Effects

Physicists have determined that main sequence stars, including our Sun, derive their energy by fusing hydrogen nuclei into helium nuclei in their super-hot core. Given the initial hydrogen-rich composition of stars in general, it's no surprise that the main sequence stage is by far the longest-lasting phase in any star's life. And, given their longevity, it follows that main sequence stars comprise the large majority of stars we observe in our galaxy. Let's see whether the data sets in Table 13.1 bear out this conclusion.

8. In Table 13.1, main sequence stars are indicated by a Roman letter "V" in the spectral type column.

   (a) Count up the number of main sequence stars in each of the two lists of Table 13.1.

(b) Use straightforward reasoning to explain why the main sequence count among the visually brightest stars is different than the main sequence count among the nearest stars.

(c) Given your answer to part 8(b), which of the two lists—visually brightest stars versus nearest stars—do you believe more accurately reflects an *unbiased* sampling of the general population of stars in our galaxy?

The star-counts in part 8 above illustrate a potentially problematic aspect of cosmic census-taking: observational selection effects. Depending on our sampling criteria—visually brightest stars, nearest stars, red stars, blue stars—a given observational database might not fully represent the stellar population we seek to study. In part 8, we find that the list of visually brightest stars is biased toward luminous giants and supergiants, which are comparatively rare, yet highly visible over great distances; that is, a considerable number of stars on this list are atypical of stars in general. By analogy, if we seek to determine the average height of human beings, it makes no sense to restrict our census only to professional basketball players. The lesson here is that astronomers must exercise care when compiling stellar databases and drawing general conclusions from them.

13.5 Theory Meets the HR Diagram

Since the 1960s, astrophysicists have programmed computers with mathematical formulations of the physical laws that govern the inner workings of stars. Into the programs they have inserted reasonable estimates of stellar mass and chemical composition. The outcome of these computational efforts is a series of center-to-surface cross-sections of interior pressure, temperature, density, and energy production for a broad array of stars. Unable to directly probe the stellar depths, astronomers have nonetheless acquired the mathematical means to deduce the physical conditions within those fiery interiors! The HR diagram is an effective way to depict the results of these computer models and compare them against the observed attributes of real stars.

Among the first generation of stellar computer models were those of newly-formed main sequence stars: gaseous protostars whose matter has completed its gravity-driven infall and commenced hydrogen-fusion energy production. Table 13.2 lists computed characteristics of these *zero-age main sequence*, or ZAMS, stars.

9. (a) In Figure 13.1 on the worksheet, plot the absolute magnitude M versus the spectral type for the ZAMS stars in Table 13.2, using the symbol "**x.**" Draw a smooth curve in the graph that best traces out the run of ZAMS data points. (b) Comment on the degree of "fit" in the HR diagram between the observed main-sequence stars and the theoretical ZAMS data. In other words, do the computer models of stars generate absolute magnitudes and

Table 13.2. Spectral Types and Absolute Magnitudes of the Zero-Age Main Sequence (ZAMS).

Spectral type	Absolute magnitude M
B0	−3.1
B5	−0.1
A0	1.6
A5	2.3
F0	3.0
F5	3.7
G0	4.5
G5	5.1
K0	6.0
K5	7.3
M0	8.9
M5	13.5

Source: Straizys, V., & Kuriliene, G. 1981, Fundamental Stellar Photometry Derived from Evolutionary Tracks, Ap&SS, 80, 353.

spectral types consistent with the observed values of these quantities? Note: The ZAMS represents hypothetical stars that have just started their main sequence phase; real main sequence stars evolve beyond the initial ZAMS state and gradually shift from their initial place in the HR diagram.

Worksheet, Activity 13: The Hertzsprung–Russell Diagram

Name _____

1.

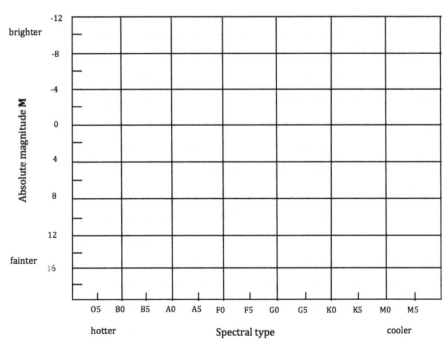

Figure 13.1. HR diagram: absolute magnitude versus spectral type.

2. (a) _____
_____

   (b) _____
_____

3. (a) _____
_____

   (b) _____
_____

4. (a) _____
_____

   (b) _____
_____

5. (a) _____
   _____
   _____

   (b) _____
   _____
   _____

6. _____
   _____
   _____

7. (a) _____
   _____
   _____

   (b) _____
   _____

8. (a) Brightest stars list: _____ main-sequence stars out of a total of _____stars
   Nearest stars list: _____ main-sequence stars out of a total of _____stars

   (b) _____
   _____
   _____

   (c) _____

9. (a) See Figure 13.1 on this worksheet.

   (b) _____
   _____

Introduction to Stars and Planets
An activities-based exploration
Alan Hirshfeld

Activity 14

The Distance to a Star Cluster

Preview

The observational and theoretical realms of astronomy combine to create a graph-based method, called main-sequence fitting, by which the distances to certain star clusters can be determined. However, the widespread presence of dust particles in space dims the incoming light of more remote clusters, requiring a correction to their distances.

14.1 The Color–Magnitude Diagram

The distances of stars within a few hundred light-years of the Sun can be measured with reasonable accuracy from the parallax method, a form of stellar triangulation based on a common land-surveying technique. Other methods must be employed to gauge the distances of stars that lie beyond the solar neighborhood. One such approach involves observation of our galaxy's *open clusters* (see Figure 14.1), each an irregularly-shaped, gravitationally-bound hive of up to several hundred stars, all presumed to be equally old. (*Globular clusters*, by contrast, are spherical in form and contain hundreds of thousands of stars.) Since open clusters come in a variety of sizes, a cluster's apparent diameter is an unreliable indicator of its distance from the Earth. However, a variant of the classic HR diagram, called a *color–magnitude diagram*, can be used to deduce a cluster's remoteness in space.

In place of the HR diagram's absolute-magnitude and spectral-type axes, a color–magnitude diagram graphs a star's apparent magnitude against a numerical measure of its color, called the *color index*, as depicted in Figure 14.2. To obtain the color index, a star's apparent magnitude is measured from the portion of its light emission that falls within two distinct wavelength bands: one in the yellow–green portion of the spectrum, designated the *visual band*, and the other in the blue portion of the spectrum. These magnitudes are represented as m_V and m_B, respectively. The color index is simply the difference $m_B - m_V$, which ranges from about -0.3 for very hot, blue stars to about $+2.0$ for very cool, red stars.

Figure 14.1. The Pleiades star cluster. (Credit: NASA, ESA, AURA/Caltech, Palomar Observatory.)

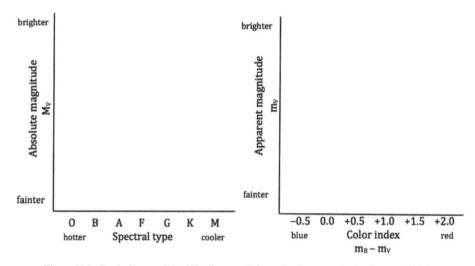

Figure 14.2. Typical axes of the HR diagram (left) and color–magnitude diagram (right).

Plotting a random selection of stars in a color–magnitude diagram yields a haphazard array of data points, since apparent magnitude takes no account of the differing distances of stars. (That's why astronomers developed absolute magnitude, which effectively situates all stars at a distance of 10 parsecs.) But star clusters present no such problem in the color–magnitude diagram: all of a cluster's stars are essentially equidistant from the Earth, assuming we neglect the relatively small volume of space over which they are spread. Thus, a cluster's main sequence—that populous diagonal band of stars in the HR diagram—will be clearly delineated in its color–magnitude diagram; red giants will sometimes appear in their usual location toward the diagram's upper right. Were they not so faint, the cluster's white dwarfs would be found toward the lower left.

14.2 Main-sequence Fitting

Let's assume that the inherent properties of zero-age main sequence (ZAMS) stars are identical everywhere in our galaxy, including from cluster to cluster. Barring alterations due to the aging of stars, the main sequences of star clusters should bear a close resemblance to each other in the color–magnitude diagram. This is borne out by observations—except for one thing. The various cluster main sequences are offset from one another in apparent magnitude, that is, along the diagram's vertical axis. The reason is simple: clusters lie at different distances; the farther ones will appear fainter. Therefore, any displacement of a cluster's main sequence from the ZAMS in the HR diagram stems from the fact that the cluster is situated farther away—often significantly farther away—than the standard 10 parsecs. The amount of this displacement, in magnitudes, can be used to determine the cluster's distance, in a process astronomers call *main-sequence fitting*.

1. Table 14.1 lists the apparent visual magnitude m_V and color index $m_B - m_V$ of a selection of stars in the Pleiades star cluster, pictured in Figure 14.1. (a) Plot this data in the color–magnitude diagram, Figure 14.3, on the worksheet. (b) Draw a line that best fits the straight-line portion of the Pleiades main sequence between color indexes 0.2 and 0.8.
2. Table 14.2 lists the absolute visual magnitude M_V and color index $m_B - m_V$ for a selection of stars on the standard main sequence, the ZAMS. (a) Plot this data in the color–magnitude diagram, Figure 14.3, on the worksheet. Here, it is permitted to graph the stars' absolute magnitudes along the apparent-magnitude axis; by definition, a star's absolute and apparent magnitude numbers are equal when the star is 10 parsecs away. For the purpose of distance estimation, we will assume that the Pleiades main sequence would coincide with the ZAMS if the cluster were moved from its present location in space to the standard distance of 10 parsecs. (b) Draw a line that best fits the straight-line portion of the ZAMS between color indexes 0.2 and 0.8.
3. (a) From your completed color–magnitude diagram, compute the magnitude difference between the Pleiades main sequence and the standard main sequence for the color index 0.2. In other words, for color index 0.2, (i) read the corresponding magnitude $m_{Pleiades}$ of the Pleiades main sequence, to the nearest tenth of a magnitude; (ii) read the corresponding magnitude m_{ZAMS}; and (iii) subtract the two magnitudes: $m_{Pleiades} - m_{ZAMS}$. Enter your answer on the worksheet. (b) Repeat this procedure for color indexes 0.4, 0.6, and 0.8.
4. Compute the average of the four magnitude-differences from part 3. (Add the four numbers and divide by four.) Write your answer on the worksheet.

Table 14.1. Apparent Visual Magnitude m_V and Color Index $m_B - m_V$ of Selected Stars in the Pleiades Cluster

Name or ID #	Apparent Magnitude m_V	Color Index $m_B - m_V$
Alcyone	2.87	−0.09
Atlas	3.64	−0.08
Electra	3.71	−0.11
Maia	3.88	−0.07
Merope	4.18	−0.06
Taygeta	4.31	−0.11
Pleione	5.09	−0.08
Celaeno	5.46	−0.04
Asterope 1	5.76	−0.04
Asterope 2	6.43	−0.02
801	6.85	0.04
153	7.51	0.15
157	7.90	0.34
232	8.06	0.20
158	8.23	0.25
697	8.60	0.35
470	8.95	0.39
25	9.47	0.48
233	9.66	0.52
403	9.83	0.54
708	10.13	0.62
102	10.51	0.71
173	10.86	0.85
248	11.02	0.77
193	11.29	0.81
129	11.47	0.88
916	11.71	0.87
34	12.03	0.94

This average represents our best estimate of the magnitude offset between the Pleiades main sequence and the ZAMS. Our working premise is that the relative dimness of the Pleiades main sequence compared to the ZAMS is entirely due to the cluster's greater distance.

5. Recall from a previous activity that a one-magnitude difference in stellar brightness represents a brightness ratio of approximately 2.5; a two-magnitude difference represents a brightness ratio of approximately $(2.5)^2$; a three-magnitude difference, $(2.5)^3$; and so on. Thus, to transform the magnitude difference between the main sequences in part 4 into a brightness ratio, compute $(2.5)^{m_{Pleiades} - m_{ZAMS}}$. (Your calculator has a key labeled y^x, or similar, that performs this operation; for practice, confirm that 2^3 yields 8 and $(2.5)^5$ is close to 100.) Enter your answer into the worksheet.

Table 14.2. Absolute Magnitude M_V and Color Index $m_B - m_V$ of Selected Stars on the Zero-age Main Sequence (ZAMS)

Absolute magnitude M_V	Color index $m_B - m_V$
0.6	−0.10
1.5	0.00
1.9	0.10
2.4	0.20
2.8	0.30
3.4	0.40
4.1	0.50
4.7	0.60
5.2	0.70
5.8	0.80
6.3	0.90
6.7	1.00
7.1	1.10
7.5	1.20
8.0	1.30
8.8	1.40
10.3	1.50

The brightness ratio you computed in part 5 specifies how many times brighter the standard ZAMS appears than the Pleiades main sequence. But it is well known that a star's brightness diminishes according to the *inverse square of its distance*: $(b_1/b_2) = (r_2/r_1)^2$, where b_1 and b_2 are the brightnesses of stars situated at distances r_1 and r_2, respectively. "Reversing" this formula, we see that the ratio of star distances diminishes according to the *inverse square root* of the brightness ratio: $(r_2/r_1) = \sqrt{b_1/b_2}$.

Therefore, the square root of the brightness ratio found in part 5 reveals how many times farther away the Pleiades cluster is than the ZAMS. Although the ZAMS is a theoretical construct, we do associate it with a specific distance: since the brightness of its stars is expressed in terms of *absolute* magnitude, that effectively places the ZAMS at the standard distance of 10 parsecs. In other words, to figure out the distance of the Pleiades, in parsecs, simply take the square root of the brightness ratio in part 5, and multiply by 10.

6. Compute the square root of the brightness ratio in part 5.
7. Multiply by 10 your answer to part 6 to determine the distance of the Pleiades star cluster, in parsecs.

14.3 Caveat: Interstellar Dust

The validity of distances deduced by this main-sequence fitting procedure hinges on a critical assumption: that the intervening space between the Earth and the star cluster is free of obscuring material. Such interstellar matter might artificially dim a cluster's main sequence stars and thereby lead to an erroneous distance. In fact, vast clouds of dust are evident in photographs of the Milky Way. A famous study by American astronomer Robert Trumpler in the 1930s revealed that many regions of our galaxy contain widely scattered dust particles which together diminish the light of star clusters. Because it lies relatively close to the Earth, the Pleiades cluster in this activity is not significantly affected by dust-dimming; however, to yield correct distances, the main-sequence brightness of more remote clusters must be corrected for obscuration caused by interstellar dust.

8. Suppose the main sequence stars of the Pleiades cluster are each dimmed by, say, one-tenth of a magnitude as their light encounters space dust on their way to the Earth. We must therefore adjust the average magnitude difference $m_{Pleiades} - m_{ZAMS}$ in part 4 to negate this dimming effect. (a) Correcting for this hypothetical 0.1 magnitude dimming, what would your adjusted average magnitude difference in part 4 become? Explain how you arrived at your answer. (b) Using your adjusted average magnitude difference in part 8(a), repeat parts 5, 6, and 7. What is your adjusted distance of the Pleiades, assuming a 0.1 magnitude dimming of its light?

9. Notice that at color indexes smaller ("bluer") than about zero, the Pleiades main sequence deviates upward from the straight-line form of the rest of its main sequence, as well as from the straight-line form of the ZAMS. Can you think of a reason why these stars no longer lie on the main sequence? Hint: The ZAMS consists of "zero-age" stars that have just commenced hydrogen fusion in their core; also, heavier stars consume their fuel more quickly than lighter-weight stars.

Worksheet, Activity 14: The Distance to a Star Cluster

Name _____

1, 2.

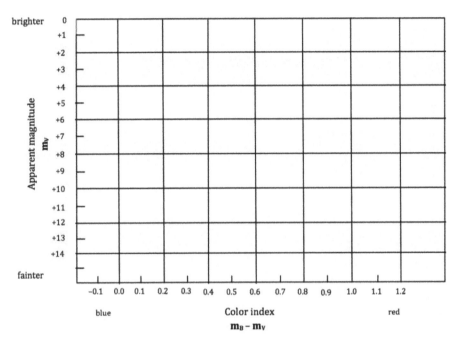

Figure 14.3. Color–magnitude diagram.

3. (a) $m_{Pleiades} - m_{ZAMS}$ at color index 0.2 = _____
   (b) $m_{Pleiades} - m_{ZAMS}$ at color index 0.4 = _____
   $m_{Pleiades} - m_{ZAMS}$ at color index 0.6 = _____
   $m_{Pleiades} - m_{ZAMS}$ at color index 0.8 = _____

4. Average magnitude difference = _____

5. Brightness ratio = _____

6. Square root of brightness ratio = _____

7. Distance of the Pleiades = _____ parsecs

8. (a) Adjusted average magnitude difference = _____
   _____
   _____

   (b) Adjusted distance of the Pleiades = _____ parsecs

9. _____
   _____
   _____
   _____

Introduction to Stars and Planets
An activities-based exploration
Alan Hirshfeld

Activity 15

The Evolution of the Sun

Preview

The Sun evolves over time, as changes in its thermonuclear energy source alter its interior structure and external appearance. This gradual evolution is revealed by computer simulations of the physical processes that govern solar matter and energy. When plotted in the Hertzsprung–Russell diagram, the computer data traces out the Sun's pathway through life as a star.

15.1 Programming the Sun

The Sun is an exemplar of the stellar species. And being the only one of its kind near enough to offer easy inspection—not to mention its role in our very existence—it is the most studied star of all. Fundamentally, the Sun is no more than a huge ball of gas, whose lifelong exertion is to ward off collapse under its own gravity. The aggregate weight of its overlying gaseous layers creates a high-density, high-pressure, high-temperature environment deep inside the Sun. Yet, the laws of physics apply within this hellish domain, as we believe they do everywhere in the universe. From our safe overlook here on Earth, we can wield these physical principles to probe the Sun's fiery interior and apply what we learn to other stars.

Over many decades, astronomers have developed computer-based simulations of the Sun—essentially, a set of programmable, mathematical formulas that stand in for the real physical processes that stabilize and generate energy inside our star. Not only do these simulations depict the Sun as it is now, but also how it probably appeared in the past and how it will likely evolve going forward. Table 15.1 contains the basic quantities that define what astronomers call the Standard Solar Model, that is, a computer-derived avatar of the present-day Sun. The complex equations used to generate the Standard Solar Model's physical attributes, which are listed in the opening row of Table 15.1, can be incremented forward in time to generate the rest of the table's information. We will use this remarkable data set to explore the Sun's evolution.

Table 15.1. Data for the Standard Solar Model

Stage of evolution	Energy source(s)	Age (billions of years)	Luminosity (solar units)	Absolute magnitude M	Surface temperature (K)	Spectral type	Radius (solar units)	Mass (solar units)
MS	H_c	0	0.70	5.13	5600	G6	0.9	1.00
MS	H_c	4.55	1.00	4.74	5800	G2	1.0	1.00
MS	H_c	7.56	1.33	4.43	5800	G2	1.1	1.00
MS	H_c	9.37	1.67	4.18	5800	G2	1.3	1.00
MS	H_c	10.91	2.21	3.88	6500	F6	1.6	1.00
MS	H_c	11.64	2.73	3.65	4900	K2	2.3	1.00
SG	H_s	12.09	17.3	1.64	4700	K4	6.4	1.00
SG	H_s	12.15	34.0	0.91	4500	K5	9.5	0.99
RG	H_s	12.23	2350	−3.69	3100	M4	170	0.72
RG	H_s	12.23	57.7	0.34	4600	K1	12	0.72
HB	$He_c + H_s$	12.23	41.0	0.71	4700	K0	9.5	0.72
HB	$He_c + H_s$	12.24	45.9	0.59	4700	K0	10	0.72
HB	$He_c + H_s$	12.32	42.4	0.67	4800	G8	9.4	0.71
HB	$He_c + H_s$	12.34	110	−0.36	4500	K2	18	0.71
HB	$He_c + H_s$	12.35	130	−0.54	4400	K2	20	0.71
AGB	$He_s + H_s$	12.365	3000	−3.95	3200	M4	180	0.59
AGB	$He_s + H_s$	12.365	5200	−4.55	3700	M0	177	0.55
PN	—	12.365	90	−0.15	74,000	A0?	0.06	0.54
WD	—	12.37	~0.01	~10	~15,000	~A5	~0.01	~0.5

Notes. Key to abbreviations: MS, main sequence; SG, subgiant; RG, red giant; HB, horizontal branch; AGB, asymptotic giant branch; PN, planetary nebula; WD, white dwarf; H_c, core hydrogen fusion; H_s, shell hydrogen fusion; He_c, core helium fusion; He_s, shell helium fusion. The surface temperature is given in Kelvin units; subtract 273 to obtain °C. A tilde (~) indicates an approximate value. Source: Data from Sackmann, I.-J., Boothroyd, A. I., and Kraemer, K. E. 1993, Our Sun. III. Present and future, ApJ, 418, 457. © 1993. The American Astronomical Society. All rights reserved.

15.2 The Life and Death of Our Star

Although the Sun is believed to have condensed from a huge interstellar gas cloud over many millions of years, astronomers generally regard the Sun's birth—age "zero"—as the moment its core became hot enough to initiate thermonuclear fusion of hydrogen nuclei into helium nuclei. As we have seen, this stage of stellar development is called the *main sequence*, a term that echoes the populous, diagonal band of stars plotted in the Hertzsprung–Russell (HR) diagram. Most of the stars we see in outer space fall into this main-sequence category.

As the Sun's core-hydrogen supply runs low—and eventually runs out—the interior and exterior of our familiar star will undergo a series of profound changes. First the Sun will brighten and enlarge into a *subgiant* star, as hydrogen-fusion transitions from the central core to a shell of gas surrounding the core. This shell-fusion process will accelerate, causing the Sun to swell into a *red giant*, about the size of the Earth's orbit. Upon reaching a threshold temperature of about 100 million degrees, helium will ignite in the Sun's core, and the Sun will shrink into an intermediate-size *horizontal branch* star. The exhaustion of helium in the core brings about a second burst of growth, into an *asymptotic giant branch* star, similar to the Sun's previous manifestation as a red giant. Throughout these bloated giant stages, the Sun will shed a considerable amount of its mass into space, and will even expel an entire atmospheric layer to create a *planetary nebula*. Eventually, all fusion will cease and the Sun will shrink into a hot, dense, Earth-sized star called a *white dwarf*. Astronomers have learned that this series of stages applies to all stars whose starting mass is similar to the Sun's.

The Sun's evolution is frequently depicted in the HR diagram as a winding path that traces out its changing light emission and surface temperature, or equivalently, its absolute magnitude and spectral type. Note that the Sun's track in the HR diagram represents its changing visual characteristics over time, not its physical movement through space.

1. On the set of axes in Figure 15.1 on the worksheet, plot the spectral type and absolute magnitude data from Table 15.1 to depict the Sun's evolutionary path through the HR diagram. Connect the data points, *in their proper chronological sequence*, with a series of curves or lines, as appropriate.
2. Based on the data in Table 15.1, label the relevant sections of your HR diagram with the evolutionary stages of a Sun-like star: main sequence (MS), subgiant (SG), red giant (RG), horizontal branch (HB), asymptotic giant branch (AGB), planetary nebula (PN), and white dwarf (WD).
3. From the data in Table 15.1, answer the following questions: (a) In which stage is the Sun's luminosity the most stable for the longest time? (b) In which stage is the Sun most luminous? (c) least luminous? (d) largest? (e) smallest? (f) hottest? (g) coolest?
4. (a) From the data in Table 15.1, determine the duration of the Sun's main sequence stage, in billions of years.
   (b) Compute the collective duration of *all* the Sun's stages after the main sequence, from subgiant onward, in billions of years.

(c) What fraction of the Sun's overall lifetime is spent in these post-main-sequence stages compared to the main sequence stage itself? That is, divide your answer to part 4(b) by your answer to part 4(a).

(d) Write down a statement that describes the relative amount of time spent by Sun-like stars in the main sequence versus time spent after the main sequence.

(e) Use your statement in part 4(d) to justify why, in a random sampling, the majority of stars will occupy the main sequence in the HR diagram.

5. The last column in Table 15.1 lists the Sun's mass at the various stages of its life. (a) Approximately what fraction of the Sun's initial mass has been expelled into space by the time the Sun reaches its final white dwarf phase? (b) Speculate on how such a significant mass loss might affect the strength of the Sun's gravity and the stability of the Earth's orbit.

Worksheet, Activity 15: The Evolution of the Sun

Name _____

1, 2.

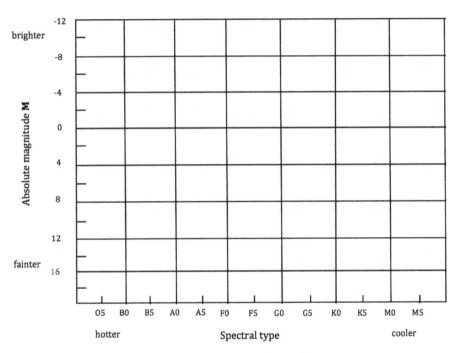

Figure 15.1. Hertzsprung–Russell (HR) diagram.

3. (a) _____
   (b) _____
   (c) _____
   (d) _____
   (e) _____
   (f) _____
   (g) _____

4. (a) _____
   (b) _____
   (c) _____
   (d) _____
      _____
   (e) _____
      _____

5. (a) _____

   (b) _____

       _____

       _____

Introduction to Stars and Planets
An activities-based exploration
Alan Hirshfeld

Activity 16

The Evolution of Massive Stars

Preview

Computer simulations of internal stellar processes indicate that massive stars evolve faster than the Sun, start at a different location on the main sequence, grow to huge dimensions, and fuse chemical elements heavier than hydrogen and helium to extend their lives. Some of these stars enter a final stage that results in a stellar explosion called a supernova.

16.1 Cosmic Beacons

Massive stars—those heavier than, say, 10 Suns—are the beacons of our galaxy. Although far less populous than "lightweight" stars like the Sun, they declare their presence by the prodigious amount of luminous energy they spew into space. The physical properties and life histories of massive stars have been studied by astronomers seeking to learn how these stellar powerhouses differ from their Sun-like cousins in their evolution and their manner of death.

Having condensed from sprawling interstellar gas clouds, massive stars begin their lives, like the Sun, on the main sequence, fusing hydrogen nuclei into helium nuclei in their core. But their greater bulk alters their internal temperature-, pressure-, and density profiles in ways that significantly affect their rate of thermonuclear fusion, modes of energy transfer, external appearance, and overall lifetime. As with the Sun and its lightweight stellar counterparts, astronomers have developed computer simulations of massive stars, dealing as best they can with the complex physics that governs these hulking members of the starry realm. While much progress has been made in the theoretical domain, much remains to be settled regarding the effects of rotation, mass loss, magnetic fields, heat-driven convection, and internal mixing on the evolution of massive stars.

Tables 16.1–16.3 list the computer-derived physical characteristics of 15, 25, and 40 solar-mass stars with an initial Sun-like chemical composition. During its lifetime, the Sun will achieve core temperatures sufficiently high to initiate the

Table 16.1. Data for the 15 Solar-mass Stellar Model

Energy Source(s)	Age (Millions of Years)	Luminosity (Solar Units)	Absolute Magnitude M	Surface Temperature (Kelvin)	Radius (Solar Units)	Mass (Solar Units)
H_c	0	20,000	−6.0	31,500	5	15.0
H_c	5.274	28,000	−6.4	30,000	6	15.0
H_c	8.669	37,000	−6.7	28,000	8	14.9
H_c	10.448	44,000	−6.9	26,000	11	14.9
H_s	12.105	57,000	−7.1	27,000	11	14.8
H_s	12.129	56,000	−7.1	10,000	84	14.8
H_s	12.131	46,000	−6.9	5000	280	14.8
$He_c + H_s$	12.132	32,000	−6.5	3500	530	14.8
$He_c + H_s$	12.351	75,000	−7.4	3300	850	14.5
$He_s + H_s$	13.738	109,000	−7.9	3300	950	13.0
$C_c + He_s + H_s$	13.754	155,000	−8.2	3300	1170	12.9
$C_s + He_s + H_s$	13.756	128,000	−8.0	3300	1100	12.9

Notes. Key to abbreviations: H_c, core-based hydrogen fusion (main sequence stage); H_s, shell-based hydrogen fusion; He_c, core-based helium fusion; He_s, shell-based helium fusion; C_c, core-based carbon fusion; C_s, shell-based carbon fusion. The surface temperature is given in Kelvin units; subtract 273 to obtain °C.
Source: Data from Maeder, A., & Meynet, G. 1987, Grids of evolutionary models of massive stars with mass loss and overshooting, A&A, 182, 243.

Table 16.2. Data for the 25 Solar-mass Stellar Model

Energy Source(s)	Age (Millions of Years)	Luminosity (Solar Units)	Absolute Magnitude M	Surface Temperature (Kelvin)	Radius (Solar Units)	Mass (Solar Units)
H_c	0	79,000	−7.5	39,000	6	25.0
H_c	2.940	110,000	−7.9	37,000	8	24.7
H_c	5.939	160,000	−8.3	30,000	15	23.8
H_c	6.897	180,000	−8.4	28,000	18	23.0
H_s	7.089	200,000	−8.5	29,000	17	22.8
H_s	7.094	210,000	−8.6	20,000	38	22.8
$He_c + H_s$	7.103	230,000	−8.7	4500	790	22.8
$He_c + H_s$	7.144	250,000	−8.8	4000	990	22.4
$He_c + H_s$	7.738	290,000	−8.9	4500	880	16.8
$He_c + H_s$	7.963	300,000	−9.0	14,000	94	15.1
$He_s + H_s$	8.261	340,000	−9.1	5500	640	14.0
$C_c + He_s + H_s$	8.267	390,000	−9.2	5500	710	14.0
$C_s + He_s + H_s$	8.268	400,000	−9.3	5500	720	14.0

Notes. Key to abbreviations: H_c, core-based hydrogen fusion (main sequence stage); H_s, shell-based hydrogen fusion; He_c, core-based helium fusion; He_s, shell-based helium fusion; C_c, core-based carbon fusion; C_s, shell-based carbon fusion. The surface temperature is given in Kelvin units; subtract 273 to obtain °C.
Source: Data from Maeder, A., & Meynet, G. 1987, Grids of evolutionary models of massive stars with mass loss and overshooting, A&A, 182, 243.

Table 16.3. Data for the 40 Solar-mass Stellar Model

Energy Source(s)	Age (Millions of Years)	Luminosity (Solar Units)	Absolute Magnitude M	Surface Temperature (Kelvin)	Radius (Solar Units)	Mass (Solar Units)
H_c	0	240,000	−8.7	45,000	8	40.0
H_c	1.938	300,000	−9.0	42,000	10	38.8
H_c	4.072	410,000	−9.3	32,000	21	35.3
H_s	4.785	480,000	−9.5	24,000	41	32.4
H_s	4.795	510,000	−9.5	7000	510	32.3
$He_c + H_s$	4.802	670,000	−9.8	6000	1230	31.8
$He_c + H_s$	4.903	710,000	−9.9	5000	1030	24.7
$He_c + H_s$	4.930	720,000	−9.9	15,000	120	24.0
$He_c + H_s$	5.002	670,000	−9.8	63,000	7	22.0
$He_c + H_s$	5.084	520,000	−9.6	130,000	1.4	19.7
$He_c + H_s$	5.286	350,000	−9.1	140,000	1.0	14.0
$He_s + H_s$	5.431	250,000	−8.8	185,000	0.5	10.0
$C_c + He_s + H_s$	5.435	300,000	−9.0	220,000	0.4	9.9
$C_s + He_s + H_s$	5.436	340,000	−9.1	270,000	0.3	9.8

Notes. Key to abbreviations: H_c, core-based hydrogen fusion (main sequence stage); H_s, shell-based hydrogen fusion; He_c, core-based helium fusion; He_s, shell-based helium fusion; C_c, core-based carbon fusion; C_s, shell-based carbon fusion. The surface temperature is given in Kelvin units; subtract 273 to obtain °C.
Source: Data from Maeder, A., & Meynet, G. 1987, Grids of evolutionary models of massive stars with mass loss and overshooting, A&A, 182, 243.

fusion of both hydrogen and helium, but not carbon or other elements; however, the cores of massive stars *do* get hot enough to ignite carbon and, in subsequent stages, the elements neon, oxygen, and silicon. It doesn't matter that the computer models used here terminate with the processing of carbon. The neon-, oxygen-, and silicon-fusing phases that follow are so brief—lasting just several years to several days!—that the outer layers of the star have no time to readjust to the ignition of a new energy source. In other words, the last line of data in each of the tables below indicates the approximate age and observable features of the star just prior to its death in a titanic explosion called a *supernova*.

1. (a) Plot the surface temperature versus absolute magnitude data from Tables 16.1–16.3, to trace the evolutionary paths of 15, 25, and 40 solar-mass stars through the Hertzsprung–Russell (HR) diagram, Figure 16.1 on the worksheet. Note: The scale of the surface temperature axis is logarithmic, a common means of graphically compressing data that spans many powers of ten; here, we assign equal axis-intervals to the numerical ranges 10^3–10^4 (1000–10,000), 10^4–10^5 (10,000–100,000), and 10^5–10^6 (100,000–1,000,000). To correctly plot the data in Figure 16.1, pay close attention to the uneven spacing of the numerical labels on the temperature axis. (b) For each of the three stellar models plotted in Figure 16.1, connect the data

points, in their proper chronological sequence, with a series of curves or lines, as appropriate.

2. Along each of the three evolutionary paths in Figure 16.1, draw tick-marks (short line segments) indicating the initiation of shell-based hydrogen fusion (label this tick-mark H_s), initiation of core-based helium fusion (label He_c), initiation of shell-based helium fusion (label He_s), initiation of core-based carbon fusion (label C_c), and initiation of shell-based carbon fusion (label C_s).

3. (a) For each of the three models, use the tabular data to compute the fraction of the star's total lifetime that it spends in the main sequence stage, fusing hydrogen into helium in its core. (b) Do these fractions confirm the general belief that the main sequence is the longest stage in a massive star's life? Explain.

4. The data tables also reveal that massive stars swell to around a *thousand* times the Sun's radius during the lowest-temperature phases of their lives. Suppose such a supergiant star were placed at the center of our solar system. Which of the solar system's inner planets would lie *within* that star? (The Sun's radius is about 0.005 au; the inner planets Mercury, Venus, Earth, and Mars are situated approximately 0.4 au, 0.7 au, 1.0 au, and 1.5 au from the Sun, respectively.)

5. (a) For each of the three models, use the tabular data to compute the fraction of the star's initial mass that is expelled into space during its lifetime. (b) Do these fractions confirm that mass loss is an important consideration in developing realistic computer models of evolving massive stars? Explain.

6. (a) For each of the three models, compute the fraction comparing the star's main sequence lifetime to that of the Sun, about 10 billion years. (b) Given that the development of intelligent life requires billions of years of stellar stability, as in the case of the Earth, would you expect to find advanced lifeforms on planets around high-mass stars? Explain.

7. Notice that the evolutionary end point of the 25 solar-mass model in the HR diagram differs markedly from that of the 40 solar-mass model. Describe this difference in terms of their respective luminosities, surface temperatures, and radii.

16.2 Heavyweights

Astronomers believe that stars whose initial mass exceeds about 40 solar-masses shed almost their entire hydrogen envelope as they evolve, revealing the super-hot layers beneath; hence, their high luminosity and bluish color. These are called Wolf–Rayet stars, and were discovered well over a century ago by their unusual emission-line spectra. (Most stars have absorption-line spectra.) The Wolf–Rayet stage lasts only a few hundred thousand years, after which the star explodes: a supernova. Stars in the 10–30 solar-mass range likewise explode, but from a cool, red-supergiant rather than a hot, blue Wolf–Rayet progenitor. To complicate matters, massive stars are often coupled in a binary-star system, in which a significant amount of gas might be exchanged between the two members as they evolve. This mass exchange process can profoundly affect the life course and ultimate fate of a star.

Worksheet, Activity 16: The Evolution of Massive Stars

Name _____

1, 2.

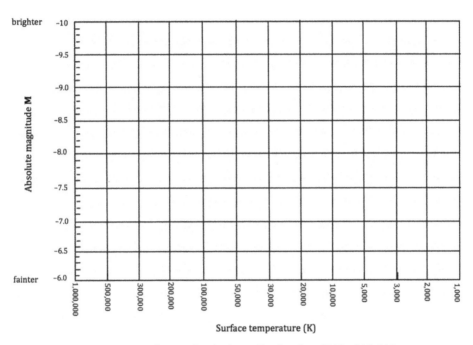

Figure 16.1. HR diagram for plotting stellar data from Tables 16.1–16.3.

3. (a) (Main sequence lifetime)/(Total lifetime) for 15 M_{solar} model: _____
   (Main sequence lifetime)/(Total lifetime) for 25 M_{solar} model: _____
   (Main sequence lifetime)/(Total lifetime) for 40 M_{solar} model: _____

   (b) _____
   _____
   _____

4. _____

5. (a) (Mass lost)/(Initial mass) for 15 M_{solar} model: _____
   (Mass lost)/(Initial mass) for 25 M_{solar} model: _____
   (Mass lost)/(Initial mass) for 40 M_{solar} model: _____

   (b) _____
   _____
   _____

6. (a) (Star MS lifetime)/(Sun MS lifetime) for 15 M_{solar} model: _____
   (Star MS lifetime)/(Sun MS lifetime) for 25 M_{solar} model: _____

(Star MS lifetime)/(Sun MS lifetime) for 40 M$_{solar}$ model: _____

(b) _____

_____

_____

7. _____

_____

_____

_____

Introduction to Stars and Planets
An activities-based exploration
Alan Hirshfeld

Activity 17

Supernovae: The Expansion of the Crab Nebula

Preview

A pair of photographs taken decades apart are used to determine the expansion rate, age, and distance of the Crab Nebula, the visible, gaseous remnant of a long-ago supernova explosion.

17.1 The Crab Nebula, Then and Now

The Crab Nebula, portrayed in Figure 17.1, is one of the most fascinating and intensively studied objects in the night sky. This rapidly expanding cloud of gaseous debris was hurled into space by a titanic supernova explosion documented during the 11th century by skywatchers in China and the Middle East. Charles Messier, famed 18th-century comet hunter, placed this ghostly apparition at the head of his now-historic list of nebulous objects, hence, its alternative designation, "M1." In the 1840s, William Parsons, the third Earl of Rosse, inspected its jagged form by eye through his 36 inch reflector telescope and dubbed it the "Crab Nebula." Photographs delivered a crisper view of its filaments and cavities, more reminiscent of exploded shrapnel than a multi-legged creature.

Modern astronomers have detected a pulsar—a pulsating radio source—at the very heart of the Crab Nebula. This ultra-dense, fast-spinning neutron star is the collapsed core of a once-massive star whose surrounding layers were hurled into space by the long-ago supernova blast. The Crab pulsar is exceptional in that it can be seen, not only by its radio-wave emission, but at optical wavelengths, as a periodically flashing pinpoint of light. By measuring the nebula's rate of expansion away from the original supernova site, marked by the pulsar, we can infer the year in which this stellar explosion took place (as seen from Earth), and compare that date to documented sightings of the event.

Figures 17.2 and 17.3 are two photographs of the Crab Nebula, the first taken in 1950 at the Palomar Observatory in California and the second a mosaic assembled from images taken in late 1999 and early 2000 by the Hubble Space Telescope. Both

Figure 17.1. Photograph of the Crab Nebula taken by the Hubble Space Telescope. (Credit: NASA, ESA and Allison Loll/Jeff Hester (Arizona State University); Acknowledgement: Davide De Martin (ESA/Hubble).)

are photographic negatives: the dark night sky is portrayed in white, while stars and glowing gases are portrayed in various shades of gray through black. This black–white reversal renders the Crab's gaseous filaments and knots more easily visible. The location of the central pulsar is indicated. The nebula's expansion during the 50 year interval separating these pictures will become evident by carefully comparing the positions of corresponding features of the nebula in each photograph relative to the fixed pattern of stars.

17.2 Image Scale

Before drawing any conclusions from the photographs, we must establish the *image scale*: the numerical relationship between measured lengths on the photographs and the corresponding angular spans in the night sky. For example, if two stars are separated by, say, 15.25 cm on the photograph, how many degrees apart are they, as viewed from Earth? The angles in this activity span tiny fractions of a degree; therefore, we will adopt the more convenient angular unit: the arcsecond, or $^1/_{3600}$ of a degree.

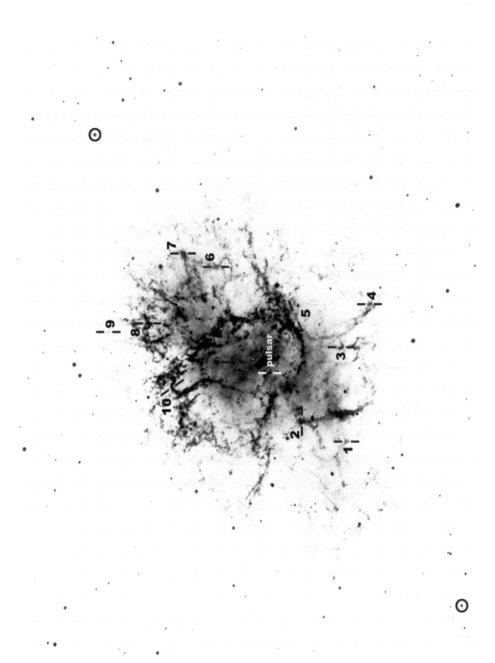

Figure 17.2. A photographic negative of the Crab Nebula taken in 1950 with the Palomar 5 meter telescope. (Credit: Palomar Observatory/California Institute of Technology.)

1. In each of the two photographs, Figures 17.2 and 17.3, use a metric-scale ruler to measure the separation between the circled pair of stars, in centimeters. In this and all subsequent measurements, you should measure to two decimal places: the tenths place of the decimal comes directly from the

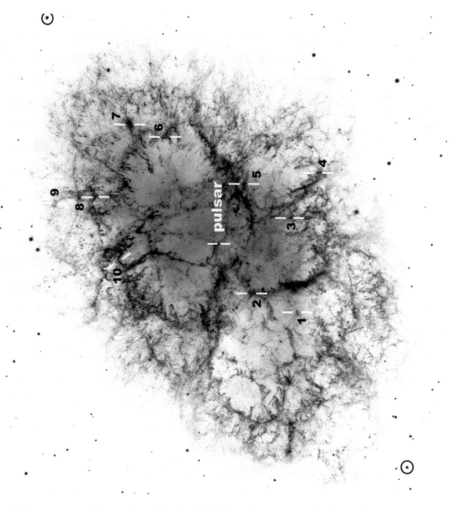

Figure 17.3. A photographic negative of the Crab Nebula taken in 2000 with the Hubble Space Telescope. (Credit: NASA, ESA and Allison Loll/Jeff Hester (Arizona State University); Acknowledgement: Davide De Martin (ESA/Hubble). CC BY 4.0.)

millimeter marks, the ruler's finest scale division, while the hundredths place comes from careful eyeballing between adjacent millimeter marks. Record the image scales of the 1950 and 2000 photographs on the worksheet. They will differ slightly; the photographs are not reproduced to the same scale.

2. The angular separation between the circled pair of stars in each photograph is known to be **468 arcsec**. For each of the two photographs, Figures 17.2 and 17.3, divide this angular separation by your corresponding linear measurement from part 1. The resulting pair of numbers are the image scales of the 1950 and 2000 photographs, respectively, in arcseconds per centimeter (arcsec cm^{-1}). Record these image scales on the worksheet.

17.3 Measurement of the Nebula

3. Ten gaseous knots have been numbered in Figures 17.2 and 17.3, each of which lies between a pair of short lines. The pulsar is similarly highlighted. For the numbered knots in the 1950 photograph (Figure 17.2), measure the separation between (i) the central pulsar, which marks the origin of the supernova explosion, and (ii) some distinctive point within the knot: its center, a protrusion, a dark patch, etc. We will refer to this center-to-knot separation as the knot's "position." Record your answers in Table 17.1 provided on the worksheet. Repeat the process for the 2000 photograph (Figure 17.3).

4. The photographic positions in part 3 are not useful for deducing the expansion rate of the nebula; they depend on the scale at which each photograph is reproduced on the page. Instead, we want each knot's *angular position*: its angular separation from the central pulsar, in units of arcseconds. The angular position depicts the actual appearance of the nebula as seen from Earth. To convert our measured positions in centimeters into angular positions in arcseconds, multiply each measurement in part 3 by the image scale for that photograph from part 2. Record the results in Table 17.1 on the worksheet.

17.4 Expansion and Age of the Crab Nebula

To simplify the analysis, we will assume that each of the supernova's gaseous filaments has traveled outward at a constant speed. Thus, the fastest knots of gas will have traveled farthest, whereas slower knots will have covered a smaller distance since the explosion. The *angular velocity* u of a knot can be computed by dividing the *change* in the angular positions of the knots in Table 17.1 by the 50 year time interval between the photographs:

$$u = \frac{\text{angular position}_{2000} - \text{angular position}_{1950}}{50}. \tag{17.1}$$

5. Using Equation (17.1), compute the angular velocity u of each knot in arcseconds per year (arcsec yr^{-1}) and record the data in Table 17.1.

6. At this point, it's straightforward to estimate the age of the Crab Nebula. We know how far each knot has traveled since the explosion—its angular position in the year 2000—and we know how fast each knot has been moving—its angular velocity u. To compute how many years T have elapsed since the explosion, divide each knot's angular position in 2000 by its angular velocity u from part 5. Enter your results in Table 17.1.

7. Don't be surprised if your answers for the elapsed time T differ considerably among the various knots; that's an inevitable consequence of any observational or experimental process, especially one like this that relies on rough measurement with a ruler. However, a more trustworthy estimate of T can be

obtained by taking the average of the 10 individual values in Table 17.1. Compute the average T_{av} of the 10 elapsed times since the explosion. (The average is computed by adding together the 10 values for T, then dividing the sum by 10.)

8. (a) Since our elapsed time estimates were based on the year 2000 photograph, subtract your T_{av} from 2000 to obtain the calendar year of the supernova that produced the Crab Nebula. (b) How does your answer compare to the established event date of 1054 A.D.? Speculate on the sources of any discrepancy between the two. Note: Given the various uncertainties and assumptions in this activity, your answer might be off by a hundred years or more.

9. While the nebula itself has expanded over the 50 year interval between the photographs, the stars recorded in these photographs are external to the nebula and remain in fixed positions relative to one another. Therefore, measurements of angular positions of stars in the two photographs should reveal no change between the years 1950 and 2000. For each of the following parts, enter your results in Table 17.2 on the worksheet.

   (a) For each photograph, measure and record the position, in centimeters, of any five stars located outside the boundary of the nebula.

   (b) Convert these positions to angular units by applying the image scale of each photograph from part 2.

   (c) Compute the change in the angular position of the stars between the two photographs, that is, subtract the angular position in 1950 from that in 2000. In general, are the changes in the angular positions of the five stars close to zero, as we might expect? If not, try to determine where the error lies and correct it.

17.5 Distance of the Crab Nebula

Measurements of features in the spectra of the Crab Nebula's luminous knots indicate that the nebula is expanding at a velocity of about **1500 km s^{-1}**. Assuming that this expansion rate has remained more or less the same since the initial blast, we can determine the Crab's overall extent in space. And by comparing this *linear diameter*—its actual width—to its *angular diameter*—its apparent width in the sky—we can deduce its distance.

10. **Linear Diameter**. The Crab Nebula's *linear diameter s* can be estimated in a few steps via the assumption that it has been expanding at the measured rate of 1500 km s^{-1} during the entire time period T_{av} that you found in part 7.

   (a) First convert T_{av} from units of years into units of seconds, given that 1 year consists of 3.15×10^7 s.

   (b) Next multiply the expansion velocity given above by the time duration, in seconds, from the previous part. The resultant number represents the *linear radius* of the Crab Nebula; multiply your answer by 2 to obtain the Crab's linear diameter, in kilometers.

(c) Finally express the Crab's linear diameter *s*, in light-years, given that there are 9.5×10^{12} km in a light-year.

11. **Angular Diameter**. Let's represent the Crab Nebula's *angular diameter* by the Greek letter "theta": θ. It is evident from the photographs that the Crab is not spherical; there is a long axis and a short axis. Let's "split the difference" between these two axes to obtain the nebula's average angular diameter, that is, how wide it might appear in the sky if it were somehow made spherical. (a) On the photograph from the year 2000, Figure 17.3, measure the approximate lengths of the Crab's long and short axes, in centimeters, then compute the average of these two numbers. (b) Multiply your answer from the previous part by the photograph's image scale, as computed in part 2. The result is the Crab's average angular diameter θ, in arcseconds.

12. **Distance**. The Crab Nebula's distance *r*, in light-years, can now be obtained using Equation (17.2), a modified version of the sector formula seen in previous activities:

$$r = 206,265 \times \frac{s}{\theta}. \tag{17.2}$$

Here *s* is the Crab's linear diameter, in light-years, from part 10; and θ is the Crab's angular diameter, in arcseconds, from part 11. (The number 206,265 arises from the angular measurement units being used.) Compare your answer to the Crab's actual distance, approximately 6500 lt-yr. If the two numbers differ by more than a few hundred light-years, try to identify the source of the discrepancy.

Worksheet, Activity 17: Supernovae: The Expansion of the Crab Nebula

Name _____

1. Separation of stars (1950): _____ cm
   Separation of stars (2000): _____ cm

2. Image scale (1950): _____ arcsec cm^{-1}
   Image scale (2000): _____ arcsec cm^{-1}

3, 4, 5, 6.

Table 17.1. Data Table for the Expansion of the Crab Nebula

Knot	Position 1950 (cm)	Position 2000 (cm)	Position 1950 (arcsec)	Position 2000 (arcsec)	u (arcsec yr^{-1})	T (yr)
1						
2						
3						
4						
5						
6						
7						
8						
9						
10						

7. T_{av} = _____ yr

8. (a) Year of explosion = _____
   (b) _____
   _____
   _____
   _____

9.

Table 17.2. Star positions in Figures 17.2 and 17.3

Star	Position 1950 (cm)	Position 2000 (cm)	Position 1950 (arcsec)	Position 2000 (arcsec)	Change in position 2000–1950
1					
2					
3					
4					
5					

10. (a) T_{av} = _____ s
    (b) Crab Nebula's linear diameter = _____ km
    (c) Crab Nebula's linear diameter s = _____ lt-yr

11. (a) Average of long and short axes on 2000 photograph = _____ cm
    (b) Crab Nebula's angular diameter θ = _____ arcsec

12. Crab Nebula's distance r = _____ lt-yr

Introduction to Stars and Planets
An activities-based exploration
Alan Hirshfeld

Activity 18

The Event Horizon of Black Holes

Preview

Principles of classical physics and Einstein's relativity theory are used to explore basic properties of black holes, objects so dense that neither matter nor light can escape their gravity. The event horizon defines the "surface" of a black hole, within which physical phenomena are hidden from external observation.

18.1 Black Hole Basics

The concept of a black hole is much older than you might think. In 1783, English cleric John Michell speculated in a letter to his scientific colleague Henry Cavendish that if the Sun's mass were sufficiently increased, its intensified gravity would prevent its own light from escaping into outer space; this ultra-heavy Sun would become dark! Although Michell's idea predates latter-day discoveries about the true nature of light and gravity, his notion of a "dark star" proved prophetic. In 1915, while serving as a volunteer in the German army during World War I, physicist Karl Schwarzschild developed the first solutions to Albert Einstein's breakthrough equations of general relativity, formulas that relate the shape of space to the distribution of matter within it. One of Schwarzschild's solutions delineated the geometry of space near a very heavy, highly concentrated mass. So warped was such a region, Schwarzschild realized, that light rays within a certain radius were trapped, creating a remarkable, non-emitting object, which astrophysicists in the 1960s dubbed a "black hole."

One route to understanding the notion of a black hole is through the intuitive concept of *escape velocity*: how fast a body must be propelled to fully free itself from the grip of a gravitating object. The greater the object's gravity, the greater the escape velocity. But according to Einstein's relativity theory, the fastest that any material body can travel through space is the speed of light. In consequence, a black hole is an object sufficiently massive that its escape velocity exceeds the speed of light; it traps everything, including light, within a boundary called the *event horizon*.

doi:10.1088/2514-3433/abc249ch18   18-1   © IOP Publishing Ltd 2020

18.2 Escape Velocity

Utilizing physical principles regarding kinetic energy (the energy of motion) and gravitational energy, physicists have derived a formula for a body's escape velocity v_{esc} from a gravitating object:

$$v_{esc} = \sqrt{\frac{2GM}{R}}, \qquad (18.1)$$

where G is a number called the universal gravitational constant (6.67×10^{-11}); M is the mass of the gravitating object, in kilograms (kg); and R is the body's distance from the center of the gravitating object—often the radius of the gravitating object itself—in meters (m). For the Earth, Equation (18.1) can be written:

$$v_{esc}(\text{Earth}) = \sqrt{\frac{2GM_E}{R_E}}. \qquad (18.2)$$

Substituting the Earth's mass and radius into Equation (18.2), we find that a body's escape velocity from the surface of the Earth is approximately 11 kilometers per second (km s^{-1}). To avoid tedious calculations, let's generate a simpler version of this escape velocity formula. Dividing Equation (18.1) by Equation (18.2) cancels out the "$2G$" terms in their respective square roots and yields a formula, Equation (18.3), in which v_{esc} is expressed as a multiple or a fraction of Earth's escape velocity of 11 km s^{-1}:

$$v_{esc\,(E)} = \sqrt{\frac{M_{(E)}}{R_{(E)}}}. \qquad (18.3)$$

Here, the subscript label "(E)" is a reminder that the mass $M_{(E)}$ and the radius (or distance) $R_{(E)}$ are to be expressed as a multiple or a fraction of the Earth's value for each of these quantities. Let's refer to escape velocities computed from Equation (18.3) as *relative escape velocities*, with the Earth's escape velocity of 11 km s^{-1} as the reference point.

Like most physics formulas, Equation (18.3) can be applied to *any* mass and radius combination, whether realistic or not. For example, we can compute the escape velocity from a galaxy's worth of matter that has been compressed to the size of a flea; yet the fact that the equation delivers a numerical answer says nothing as to whether such an object exists. Mathematical formulas are all-embracing when it comes to the range of their variables; they take in numbers and generate a result. It's the scientist's job to discover the natural laws that constrain these variables and thereby judge which mathematical results represent flights of imagination versus actual physical phenomena.

1. Using Equation (18.3), compute the relative escape velocity of the gravitating objects listed in Table 18.1 on the worksheet. Enter your answers in the table.

2. Multiply the relative escape velocity of each object in Table 18.1 by the Earth's escape velocity of 11 km s^{-1} to obtain the object's actual escape velocity, in km s^{-1}.

3. According to the physical data in Table 18.1, the mass of the Sun and the mass of a white dwarf star are roughly equivalent. Explain why their escape velocities are so different.

4. Considering your answers to parts 1 and 2, which of the escape velocities do you think are achievable using current launch and space-travel technology? Consult reference sources, if necessary.

5. A neutron star is an ultra-collapsed stellar core that packs a Sun's mass within a sphere some 10 km across. According to your answer in Table 18.1, what fraction of the speed of light (3×10^5 km s^{-1}) is the actual escape velocity from the surface of a neutron star?

18.3 John Michell's "Dark Star"

In his 1783 letter to Henry Cavendish, John Michell writes, "If the [radius] of a sphere of the same density as the Sun in the proportion of five hundred to one, and by supposing light to be attracted by the same force in proportion to its [mass] with other bodies, all light emitted from such a body would be made to return towards it, by its own proper gravity." Simply stated, Michell claims that if the Sun's present radius were extended 500 fold and if the expanded solar volume were then filled with matter at the Sun's current average density, the escape velocity from the surface of that bloated star would equal the speed of light. Let's check whether Michell's centuries-past assertion is correct.

First, we recast Equation (18.3) in solar units instead of Earth units:

$$v_{\text{esc (S)}} = \sqrt{\frac{M_{(S)}}{R_{(S)}}}. \tag{18.4}$$

Here the subscript "(S)" alerts us to the fact that the escape velocity, mass, and radius of the gravitating object are expressed in solar units, that is, as multiples or fractions of the Sun's values for these quantities. (The Sun itself is characterized by $v_{\text{esc(S)}} = 1$, $R_{(S)} = 1$, and $M_{(S)} = 1$.) In solar units, Michell's hypothetical "dark star" has a radius $R_{(S)}$ of 500, as stated in his letter.

To find the mass $M_{(S)}$ in Equation (18.4), we note that the mass of a star can be quantified by multiplying the star's *average matter density* by its *volume*. Michell informs us in his letter that the average matter density of his super-sized star is no different than that of the present-day Sun. Therefore, it's only his star's greater volume of matter that determines its overall mass in solar units. The volume of a spherical star of radius r is given by the expression $(^4/_3)\pi r^3$. Thus, if a star's radius is doubled, its volume grows by a factor of $(2)^3$, or 8 times; if the radius is tripled, the volume goes up by $(3)^3$, or 27 times; and so on. Michell's star has a radius of 500 Suns, and its volume, hence, its overall mass $M_{(S)}$, must be $(500)^3$, or **1.25×10^8 solar masses.**

6. Using Equation (18.4) with the numerical values derived above for $R_{(S)}$ and $M_{(S)}$, compute the relative escape velocity, in solar units, of Michell's theorized light-trapping star.

7. (a) Compute the escape velocity, in km s^{-1}, of Michell's star by multiplying your answer to part 6 by the actual escape velocity of the Sun in Table 18.1. (b) Does the result of the previous part suggest that Michell was correct in his reasoning about transforming the Sun (hypothetically!) into a black hole? Explain your answer. Note: In Michell's time, the speed of light was known only approximately; therefore, his adoption of a stellar radius of 500 Suns should be considered a rough estimate.

18.4 The Event Horizon

According to Einstein's theory of relativity, no material body can move through space faster than the speed of light. Therefore, a gravitating object whose escape velocity equals or exceeds the speed of light can trap matter or light rays that stray too close. The size of such a black hole can be determined by setting the escape velocity in Equation (18.1) equal to the speed of light, commonly represented by the letter c. Solving this equation for the radius variable yields:

$$R_{EH} = \frac{2GM}{c^2}. \tag{18.5}$$

Equation (18.5) is the formula for the radius R_{EH} of a black hole's event horizon— effectively, its invisible outer boundary—also known as the *Schwarzschild radius*. At the Schwarzschild radius, the escape velocity equals the speed of light; within this radius, the escape velocity exceeds the speed of light. Thus, any matter or light located within the Schwarzschild radius cannot escape from the gravitating object. As its name suggests, the event horizon prevents us from viewing any phenomenon that occurs within its bounds.

Equation (18.5) can be simplified for computation by substituting the numerical values for the constants G and c:

$$R_{EH} = 1.48 \times 10^{-30}M, \tag{18.6}$$

where M is the mass of the gravitating object, in kilograms (kg), and the resultant R_{EH} comes out in units of kilometers (km).

8. Use Equation (18.6) to compute the Schwarzschild radius R_{EH} and corresponding diameter D_{EH} of the Sun, whose mass is given in Table 18.2. This diameter represents the size within which you would have to compress all of the Sun's matter to turn it into a black hole. Enter your answer into Table 18.2 on the worksheet. Note: Given our current understanding of stellar physics, the Sun will not collapse into a black hole during its active lifetime or upon its death.

Notice that the numerical result of Equation (18.6) depends only of the mass of a given object. Since we've already computed the Sun's Schwarzschild radius, we can dispense with Equation (18.6) and instead recognize a simple multiplication rule: an object of M solar masses will have M times the Sun's Schwarzschild radius R_{EH} **(Sun)**:

$$R_{EH}(\text{km}) = M(\text{solar masses}) \times R_{EH}(\textbf{Sun}), \qquad (18.7)$$

where R_{EH} **(Sun)** has the value, in kilometers, that you computed in part 8.

9. Table 18.2 on the worksheet lists the masses, *relative to that of the Sun*, of the Earth, a typical massive star, and a supermassive black hole like the one that exists at the center of our galaxy. Use Equation (18.7) to estimate the Schwarzschild radius R_{EH} and corresponding diameter D_{EH} for each of the listed objects. Enter your answers into Table 18.2.

10. (a) Convert your answer for the Earth's Schwarzschild radius and diameter from kilometers into millimeters. (There are 1000 m in a kilometer and 1000 mm in a meter.) (b) Name an everyday object that is approximately the size of the Earth's Schwarzschild diameter. Now imagine the density of matter were the entire Earth crushed into this tiny space. That's the matter concentration required to turn the Earth into a black hole!

Worksheet, Activity 18: The Event Horizon of Black Holes

Name _____

1, 2.

Table 18.1. Mass, radius, and relative and actual escape velocities of various celestial objects

Object	$M_{(E)}$	$R_{(E)}$	$v_{esc\,(E)}$	v_{esc} (km s^{-1})
Moon	0.012	0.25		
Mars	0.1	0.5		
Sun	3×10^5	100		
Solar system (at Neptune)	3×10^5	7×10^5		
White dwarf star	3×10^5	1		
Neutron star	6×10^5	0.001		

3. _____
_____
_____

4. _____
_____
_____

5. Fraction of speed of light = _____

6. Relative escape velocity = _____

7. (a) Escape velocity = _____ km s^{-1}

(b) _____
_____

8, 9.

Table 18.2. Mass and Schwarzschild radius of various celestial objects

Object	Mass (kg)	Mass (solar masses)	R_{EH} (km)	D_{EH} (km)
Sun	2×10^{30}	1		
Earth	—	3×10^{-6}		
Massive star	—	10		
Supermassive black hole	—	4×10^{6}		

10. (a) R_{EH} **(Earth)** = _____ mm

       D_{EH} **(Earth)** = _____ mm

(b) _____

Introduction to Stars and Planets
An activities-based exploration
Alan Hirshfeld

Activity 19

Kepler's Third Law and the Masses of Black Holes

Preview

Kepler's third law of orbital motion is used to deduce the masses of black holes in binary-star systems as well as the supermassive black hole occupying the center of our Milky Way galaxy. Merging black holes generate gravitational waves whose detection provides clues to the mass and size of these unusual objects.

19.1 Introduction

In the early 1600s, half a century after Copernicus's heliocentric cosmos burst onto the scientific scene, the brilliant German mathematician Johannes Kepler undertook an analysis of planetary movements. That lengthy study led to the discovery of a set of fundamental laws to which planets adhere as they orbit the Sun. The first of these laws holds that planetary orbits are not circular, as had long been believed, but elliptical, with the Sun located slightly off-center at a point called the *focus*; thus, a planet's distance from the Sun changes as it makes its orbital rounds. Secondly, planets move faster along their orbit when closer to the Sun and slower when farther away, their speed at every instant governed by a rigorous mathematical rule.

Kepler's planetary data disclosed to him yet a third law, this one a mysterious numerical link between two orbital parameters: the orbital period P, or the time it takes the planet to circuit the Sun; and the semi-major axis a, defined as one-half of the longest dimension of the orbital ellipse. Specifically, the square of the orbital period is proportional to the cube of the semi-major axis. In equation form, Kepler's third law looks like this:

$$M = \frac{a^3}{P^2}.$$

(19.1)

This simplified version of Kepler's third law holds whenever all three variables are expressed in units relative to the Sun and the Earth. Here, M represents the mass of the central object, which for the solar system is the Sun; in solar units, the Sun has a mass of precisely 1. The semi-major axis a is expressed in astronomical units, or au, and can be taken as the average distance between the Earth and the Sun. (In fact, planetary ellipses are ever so slightly out-of-round, so the semi-major axis is essentially the radius of a comparably sized circular orbit.) The orbital period P in Equation (19.1) is measured in Earth-years; thus $P = 1$ for the Earth.

19.2 Inside the Solar System

1. As we saw in a previous activity, the orbital data for any one of the solar system's eight planets can be used in Kepler's third law to derive the Sun's mass. For example, Jupiter's orbital radius a is 5.2 au and its orbital period P is 11.86 yr. Apply this data to Kepler's third law, Equation (19.1), to confirm that the Sun's mass is indeed 1 solar mass.
2. Suppose the Sun were replaced by a black hole 10 times as massive as the Sun. How long would it take the Earth to orbit the black hole at its present orbital radius of 1 au? Express your answer in both years and days.
3. The Sun is 1.4×10^6 km across and spans an angle of about ½ degree, as seen from the Earth. The 10 solar-mass black hole in part 2 would have an event horizon diameter of about 100 km. If the event horizon were somehow visible from the Earth, how wide would it appear, in degrees? Explain how you got your answer.

19.3 Beyond the Solar System

As proven by 19th-century astronomers, Kepler's third law, Equation (19.1), applies not just to our solar system, but to any star system. For a binary star—two stars orbiting each other—we can use measurements of the orbital period and semi-major axis to compute the sum of the stars' masses. In such cases, where there is no centralized mass, the semi-major axis a is gauged by the average separation between the member stars.

Often one member of a binary star system is not visible through a telescope; it might be a very dim white dwarf star or perhaps an even smaller and denser neutron star. According to theoretical analysis, a neutron star will collapse into a black hole if it is heavier than 3 times the Sun's mass. So if Kepler's third law reveals that the unseen member of a binary system has a mass greater than 3 solar masses, that imperceptible body is most likely a black hole.

4. Based on periodic oscillations of its velocity, the cosmic X-ray source Cygnus X-1 is a binary system: a blue supergiant star paired with an unseen companion. The two objects orbit each other in a mere 5.6 days (0.0153 yr) and are separated by an estimated 0.2 au, much closer than Mercury is to the Sun. (a) Solve Kepler's third law, Equation (19.1) to find the combined mass M of the supergiant star and its unseen companion, in solar masses. (b) Spectroscopic studies indicate that the supergiant star itself has a mass of

about 20 solar masses. Use this information to deduce the mass of the unseen companion object. Is this object likely a black hole? Explain.

Astronomers believe that so-called "supermassive" black holes exist at the center of most galaxies, including our own. A supermassive black hole is far larger and heavier than a black hole produced by the collapse of a single star, yet it would be invisible in a telescope. Nevertheless, Kepler's third law gives us the means to estimate its mass: all it takes is the discovery of one star caught in the black hole's gravitational grip. Such a star has been found that orbits 1000 au from the center of our Milky Way galaxy—25 times Pluto's distance from the Sun—yet takes only 15 years to zip around its orbit. This works out to a blistering space velocity of almost 2000 km s^{-1}!

5. Apply Kepler's third law to find the combined mass M of the star and the presumed supermassive black hole described above, in units of solar masses. Note: The star's mass has been estimated at 15 solar masses; therefore, your answer for M is effectively the mass of the black hole itself. You see why it's called a "supermassive" black hole!

19.4 Beyond the Galaxy

In early 2016, astronomers at the Laser Interferometer Gravitational Wave Observatory (LIGO) announced the detection of gravitational waves: tiny ripples in the invisible underlying structure of space, whose existence was proposed a century earlier by Albert Einstein. Subsequent analysis revealed that the source of the waves was a pair of in-spiraling black holes located in a faraway galaxy, as depicted in Figure 19.1. The brief sequence of gravitational oscillations terminated abruptly in what LIGO scientists presume was the merger of the black holes. Interpretation of the data indicates that, when the detected signal was at its strongest, the centers of the two objects were spaced about 350 km apart and circled each other a dizzying 75 times a second, that is, each had an orbital period of only 0.01333 s. (Take a moment to envision this remarkable pair.) Furthermore, the form and regularity of the gravitational oscillations indicated that, prior to their merger, the black holes moved in accord with Kepler's third law. Thus, we can apply Equation (19.1) to determine the total mass of this groundbreaking black hole duo.

6. (a) Equation (19.1) requires that the separation a between the black holes—350 km—be expressed in au. Convert a from kilometers into au, given that 1 au = 1.5 × 10^8 km. (b) Equation (19.1) also requires that the orbital period P be expressed in years. Convert P from seconds into years, given that 1 yr = 3.15 × 10^7 s.
7. (a) Use Equation (19.1) with the data from part 6 to compute the total mass of the black hole pair described above. (b) LIGO astronomers concluded that the black holes had masses of 36 and 29 solar masses, respectively. Does your answer to part 7(a) conform to this result? If not, check your calculations.

Let's reflect on the fact that we are dealing here with a pair of very massive celestial objects—each as heavy as some 30 Suns—speeding along in their orbits

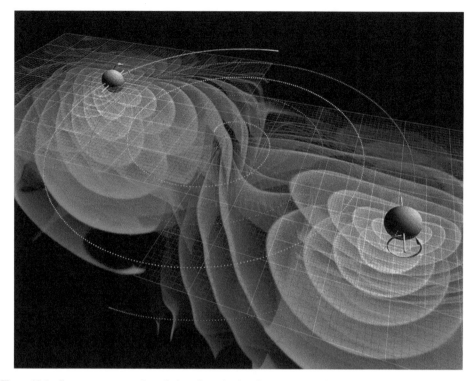

Figure 19.1. Computer-generated rendering of gravitational waves (colored contours) emitted by a pair of in-spiraling black holes. (Credit: C. Henze/NASA Ames Research Center.)

while their centers are separated by only 350 km. Although they would ultimately merge, the two bodies were still intact and distinct from one another at that point in time. Therefore, neither object could have been more than 175 km in radius (one-half of 350 km); otherwise their surfaces would have been touching and noticeably impeding their movement. The orbiting objects could not have been ordinary stars or Earth-sized white dwarfs, both of which far exceed this radius constraint. Neutron stars, which do meet the radius limit, are eliminated because, as we learned above, their mass tops out at about 3 Suns. We are driven to the conclusion that these gravity-wave emitters *must* have been black holes. But just to be sure...

8. Use Equation (19.2), introduced in the previous activity as Equation (18.7), to confirm that the event horizon radius of a 30 solar-mass black hole is indeed smaller than 175 km. (Reminder: The event horizon radius of the Sun, R_{EH} **(Sun)**, is about 3 km.)

$$R_{\text{EH}}(\text{km}) = M(\text{solar masses}) \times R_{\text{EH}}(\textbf{Sun}). \tag{19.2}$$

Worksheet, Activity 19: Kepler's Third Law and the Masses of Black Holes

Name _____

1. Sun's mass (in solar units) = _____

2. Orbital period = _____ yr = _____ days

3. Event horizon appears _____ degrees across.

4. (a) M = _____ solar masses

   (b) _____
_____

5. M = _____solar masses

6. (a) a = _____ au

   (b) P = _____ yr

7. (a) Total mass = _____ solar masses

   (b) _____

8. R_{EH} = _____ km

Introduction to Stars and Planets
An activities-based exploration
Alan Hirshfeld

Activity 20

Our Place in the Galaxy

Preview

The space distribution of globular star clusters shows a strong concentration in the direction of the galactic center. Many are located outside the dusty, light-absorbing galactic plane, allowing astronomers to view them to great distances. Thus, globular star clusters provide a means to determine the size of our Milky Way galaxy as well as the location of the solar system within it.

20.1 Our View of the Galaxy

Humanity's long-running effort to locate its place within the wider universe accelerated following the Copernican Revolution in the mid-16th century. After considerable intellectual struggle, astronomers accepted the Sun's centrality to what they sensibly dubbed the solar system. But with our cosmic horizons rolled back by the advent of the telescope and later by the introduction of celestial photography, questions arose: where is the solar system situated in space relative to the many thousands of stars visible in the night sky? And what is the layout of this galactic-scale star system? To deduce the form and extent of the Galaxy from our Earthbound vantage point proved a challenge. A single, space-age snapshot reveals the Earth's shape and size; however, no such panoramic portrait of our Galaxy is possible, constrained as we are to view it only from within.

The noted 18th-century astronomer William Herschel tried to assess the dimensions of our Galaxy by counting stars of various brightnesses in different parts of the sky. The result was a ragged-edged, roughly disk-shaped aggregation of stars many times wider in extent than the solar system. Follow-up star counts during the early 20th century implied that the Galaxy is relatively compact—perhaps 20,000 lt-yr across—and that the solar system lies at or near its center. The solar system's seemingly privileged position, with its echo of the long-discredited geocentric cosmos, troubled many astronomers, who intuited that the farther reaches of the Galaxy's disk might be hidden from view by the presence of diffuse, light-absorbing matter in interstellar space.

doi:10.1088/2514-3433/abc249ch20

© IOP Publishing Ltd 2020

20.2 Globular Star Clusters

In charting the Milky Way's size and our place within it, astronomer Harlow Shapley, at the Mount Wilson Observatory in California, expressed his belief that globular star clusters are "beacon lights point[ing] the way to the center of the Galaxy and to its edges." Figure 20.1 is a photograph of a globular cluster, number 80 in the catalog of 18th-century French comet hunter Charles Messier. By Shapley's time, dozens of these populous stellar hives had been discovered. Significantly, most of them lay above or below the presumably dusty plane of our Galaxy. As such, they would be less subject to the obscuring effect of intervening matter, Shapley asserted, and their overall distribution would therefore be an accurate gauge of the Milky Way's dimensions. (Globular clusters, which contain some of the Galaxy's oldest stars, differ from open clusters like the Pleiades, which are relatively young and tend to hug the galactic plane.)

A century before Shapley, William Herschel's son John had mapped the sky positions of globular clusters and found their distribution to be highly skewed: the vast majority of these clusters lie in one hemisphere of the sky, with fully one-third confined to a region around the constellation Sagittarius that comprises just 2% of the global night sky. Given this stark asymmetry, Shapley reasoned that one of the following conclusions must be true: (i) the Milky Way is small and centered on the solar system, with the globular clusters oddly offset to one side; or (ii) the Milky Way is large and concentric with the globular clusters, and it is our solar system that lies in the Galaxy's outskirts. In 1919, Shapley announced his observational verdict that the second of these scenarios is true.

Shapley and his successors deduced the distance to individual globular clusters by various means; for the most part, they observed specific categories of stars within a cluster whose absolute magnitudes had been previously determined from other

Figure 20.1. Globular star cluster Omega Centauri, one of approximately 160 such clusters discovered to date in the Milky Way galaxy. (Credit: La Silla Observatory/ESO. CC BY 4.0.)

studies. Table 20.1 lists the distances, in thousands of light-years, for a sample of 20 globular clusters that lie above or below the dust-infused galactic plane. (Approximately 160 globular clusters have been discovered in and around our Galaxy; several of the brighter ones are easily visible through a small telescope.) The distances in Table 20.1 have been corrected for the light-dimming effect of interstellar dust, which is minor for most of the clusters listed here.

20.3 Galactic Coordinate System

Astronomers have devised a special, three-dimensional coordinate system to specify the locations of globular clusters in the galactic frame of reference. The system has its origin at the solar system and consists of three coordinate axes designated **X**, **Y**, and **Z**, as depicted in Figure 20.2.

The three galactic coordinate system axes are defined as follows:
- The **X**-axis lies within the galactic plane. Its values are positive toward the galactic center (in the constellation Sagittarius) and negative away from the galactic center. Thus, a positive value of **X** indicates that an object lies closer to the galactic center than the Earth does, a negative value farther away.

Table 20.1. Globular Cluster Data

Cluster	Distance (×1000 lt-yr)	X (×1000 lt-yr)	Y (×1000 lt-yr)	Z (×1000 lt-yr)
47 Tucanae	15	6	−8	−10
NGC 288	29	0	0	−29
NGC 2298	35	−14	−31	−10
M 68	34	13	−23	20
NGC 5466	52	11	10	50
IC 4499	61	35	−46	−22
NGC 5824	105	86	−45	39
Palomar 5	76	53	1	54
NGC 5897	41	34	−10	21
M 5	24	17	1	18
M 80	33	31	−4	11
M 13	23	9	15	15
NGC 6356	49	48	6	9
M 54	86	83	8	−21
NGC 6723	28	27	0	−8
M 75	68	57	21	−30
M 72	55	38	27	−30
NGC 7006	134	56	113	−45
M 15	34	13	27	−16
M 30	26	16	8	−19

Notes. Abbreviations: M: Messier's list; NGC: New General Catalogue; IC: Index Catalogue. Source: Data from Harris, W. E. 1996, Catalog of parameters for Milky Way globular clusters, AJ, 112, 1487. © 1996. The American Astronomical Society. All rights reserved; 2010 edition, http://physwww.mcmaster.ca/~harris/mwgc.dat.

Figure 20.2. The galactic (**X, Y, Z**) coordinate system superimposed on an image of the spiral galaxy UGC 2885. Were this our Milky Way Galaxy, the solar system would be located at the origin of the coordinate system where the arrows' bases meet. The bright object at right is a foreground star. (Credit: NASA; ESA; B. Holwerda, U. of Louisville. CC BY 4.0.)

- The **Y**-axis also lies within the galactic plane, perpendicular to the **X**-axis. Its values are positive in the direction of the Galaxy's rotation (toward the constellation Cygnus) and negative the opposite way.
- The **Z**-axis is oriented perpendicular to the galactic plane, positive toward the north galactic pole and negative toward the south galactic pole. The **Z** coordinate indicates how far above or below the galactic plane a celestial object lies. **Z** = 0 corresponds to an object within the galactic plane.

Table 20.1 lists the galactic coordinates (**X, Y, Z**), each in thousands of light-years, of the 20 globular clusters chosen for this activity. Using this data, we can recreate Harlow Shapley's groundbreaking determination of the location of the galactic center. Figure 20.3 on the worksheet is a pair of axes on which we can plot the location of each globular cluster according to its **X** and **Z** coordinates. When completed, this graph will depict an **X–Z** cross-section of the Galaxy's globular cluster distribution, that is, how the arrangement of clusters would appear to an observer located outside the Galaxy, but in the same plane as the galactic disk.

1. Plot the **X** and **Z** coordinates of each globular cluster in Table 20.1 on the axes in Figure 20.3 on the worksheet.
2. Estimate by eye the central point (the "locus") around which the globular clusters are most evenly distributed, that is, about as many globular clusters lie above this point as below, and about as many lie to the left of this point as to the right. Mark this central point with an "**x**". According to Shapley's hypothesis, this point marks the galactic center.
3. How many light-years from the galactic center is our solar system, according to Figure 20.3?
4. As a check on your graphical estimation of the distance to the galactic center, compute the numerical average of all the **X** coordinates in Table 20.1. The result should be close to your previous eyeball estimate in part 3.
5. Modern-day observations indicate that the galactic disk extends as far as 50,000 lt-yr from the galactic center and is about 2000 lt-yr thick.

    (a) Draw a pair of short vertical lines on the **X**-axis of Figure 20.3 that mark the outer boundaries of the galactic disk in the positive-**X** and negative-**X** directions, respectively.

    (b) Now draw a pair of short horizontal lines on the **Z**-axis that mark the upper and lower boundaries of the galactic disk, respectively. Extend these horizontal lines to the right and to the left until they reach the vertical lines you drew in the previous part.

    (c) Our Galaxy has a roughly spheroidal bulge in its central region whose height is several times the thickness of the galactic disk. To represent the presence of the bulge and complete the galactic profile, trace the outline of a small coin or your fingertip around the central "**x**" you drew in Figure 20.3.

6. What fraction of the Galaxy's disk-radius is the Sun's distance from the galactic center? In other words, determine the value of the fraction in the following statement: the solar system lies [*fraction*] of the way from the galactic center to the Galaxy's outermost edge.

7. Now that we've situated our solar system within the Galaxy, we can determine approximately how fast we are orbiting around the galactic center. Express your answers to the following in powers-of-ten notation.

    (a) Adopting your answer to part 3 for the radius R of our solar system's orbit, compute the circumference C of the solar system's galactic orbit (which we presume is a circle) in light-years, where $C = 2\pi R$.

    (b) Convert your answer in the previous part from light-years into kilometers, given that there are roughly 9×10^{12} km in a light-year.

    (c) Astronomers estimate that the Galaxy rotates once every 200 million years or so at the solar-system's distance from the galactic center. To obtain the solar system's velocity V around its galactic course, in kilometers per hour, divide your answer to the previous part by the solar system's orbital period, *expressed in hours*. (There are 8760 hr in a year.)

Right now, as you sit pondering this activity, you are simultaneously circling the Earth's axis, careening around our central Sun, and circumnavigating the Galaxy. Although we inhabit the universe, nature veils its relentless motions from our senses.

Worksheet, Activity 20: Our Place in the Galaxy

Name _____

1, 2.

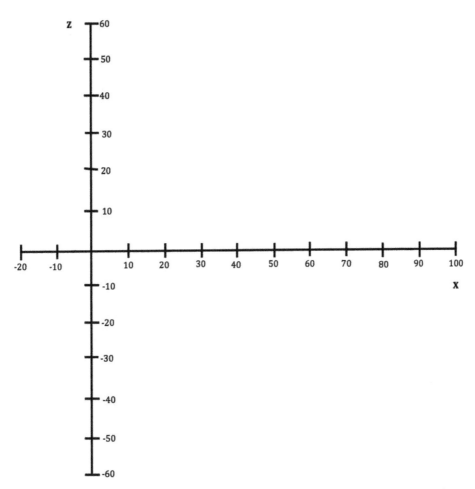

Figure 20.3. X–Z coordinate axes for globular cluster data in Table 20.1. The **X**- and **Z**-axis numbers are expressed in thousands of light-years.

3. _____ lt-yr

4. _____ lt-yr

5. (a), (b), (c) See Figure 20.3.

6. Fraction = _____

7. (a) $C = $ _____ lt-yr

(b) $C = $ _____ km

(c) $V = $ _____ km hr^{-1}

Part III

The Planets

Introduction to Stars and Planets
An activities-based exploration
Alan Hirshfeld

Activity 21

A Slice of Earth

Preview

A scaled-down rendering of the Earth presents a more accessible perspective on the depths and heights of various features of the Earth's interior, surface, and atmosphere.

21.1 The Earth in Perspective

To living creatures inhabiting its surface and its ocean depths, the Earth is a big place. Spanning more than 12,000 km (8000 miles), the spherical shape of our home planet is practically imperceptible from our low-level vantage point, yet plainly visible from space as shown in the satellite-derived image Figure 21.1. From scientific measurements and theory, scientists have learned a great deal about the Earth's geological features, atmospheric system, and interior structure. From surface to center, the Earth's interior is roughly divided into concentric regions of increasing density: crust, mantle, liquid outer core, and solid inner core. At their deepest, oceans plunge more than 10 km (6 miles) below sea level. At their highest, mountains poke upwards of 8 km (12 miles) into the air. The atmosphere that separates us from the deadly vacuum of outer space becomes exceedingly thin above 1000 km (600 miles).

The measurement units cited here—kilometers and miles—provide only a partial sense of the depth, height, or extent of the Earth's structural components: from experience, we might have an intuitive sense of the length of a kilometer or a mile, but 1000 km or 600 miles is more difficult to envision. In dealing with astronomical systems, it helps to create a reduced-scale model, that is, shrink every component of the system by the identical factor so we can view them all at once.

Figure 21.2 on the worksheet depicts a "pie slice" of a cross-section of the spherical Earth. A metric ruler (one that has millimeter markings) will be used to make measurements of various key features within or above this scaled-down segment of our planet.

doi:10.1088/2514-3433/abc249ch21

© IOP Publishing Ltd 2020

Figure 21.1. True-color image shows North and South America as they would appear from space 35,000 km (22,000 miles) above the Earth. (Credit: Reto Stöckli, Nazmi El Saleous, Marit Jentoft-Nilsen, NASA/GSFC. NASA Earth Observatory. CC BY 2.0.)

1. (a) Using the metric ruler plus the fact that the Earth's radius is 6400 km, determine the scale of Figure 21.2 in kilometers per millimeter (km mm^{-1}). (b) Explain how you arrived at your answer.

2. Adopting the km mm^{-1} scale from part 1(a), enter the number of millimeters corresponding to each of the kilometer measures of the objects or landmarks listed on the worksheet. Mark and label the location of each object or landmark, *to the proper scale*, on the cross-section of Earth in Figure 21.2. If it is not possible to visibly represent a given object or landmark to its proper scale, write "not possible" below the corresponding description on the worksheet.

Worksheet, Activity 21: A Slice of Earth

Name _____

1.

    (a) Scale of Figure 21.2 = _____ km mm^{-1}

    (b) _____

        _____

2. (a) Radius of Earth's solid inner core: 1100 km = _____ mm from Earth's center

    (b) Radius of Earth's liquid outer core: 3400 km = _____ mm from Earth's center

    (c) Typical depth of a volcano's lava source: 100 km = _____ mm below Earth's surface

    (d) Typical depth (thickness) of Earth's continental crust: 30 km = _____ mm below Earth's surface

    (e) Deepest ocean trench (Marianas Trench): 11 km = _____ mm below Earth's surface

    (f) World's deepest mine shaft: 4 km = _____ mm below Earth's surface

    (g) Tallest mountain (Mt. Everest): 8.8 km = _____ mm above Earth's surface

    (h) Altitude of the highest clouds: 7 km = _____ mm above Earth's surface

    (i) 99% of the atmosphere's mass lies below this altitude: 30 km = _____ mm above Earth's surface

    (j) "Top" of the atmosphere: 1300 km = _____ mm above Earth's surface

    (k) Altitude of the International Space Station: 400 km = _____ mm above Earth's surface

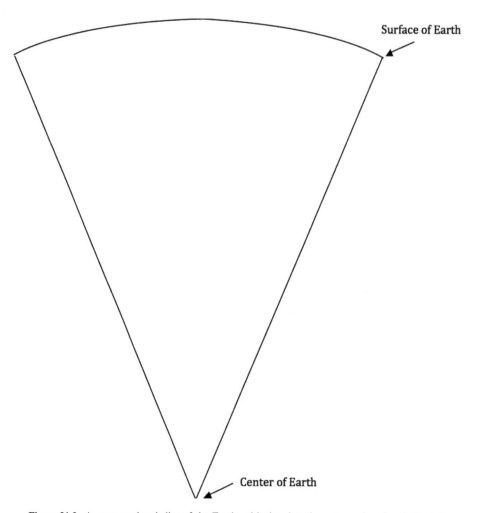

Figure 21.2. A cross-sectional slice of the Earth, with the planet's center and surface indicated.

Introduction to Stars and Planets
An activities-based exploration
Alan Hirshfeld

Activity 22

Geological Time in Perspective

Preview

During its long lifetime, the Earth has gone through significant changes, both within and on or above its surface. A compressed model of the passage of time is used to place these events into a familiar chronological context, that of a calendar.

22.1 Cosmic Calendar

The Earth is estimated to be about 4.6 billion years old. Momentous events took place during this lengthy stretch of time, such as the onset of life, creation of the Earth's atmosphere, extinction of the dinosaurs, rise of civilization, and so on. But 4.6 billion, or 4.6×10^9, years is too long a time span to relate to everyday experience, so let's scale down the Earth's age to a more accessible number: one calendar year. That is, let's squeeze Earth's entire 4.6 billion-year history into 365 days, starting with the Earth's formation at midnight on January 1 and ending with our present-day world just before the clock strikes midnight on December 31. Every past occurrence, whether long-ago or recent, will map to a specific day—perhaps to a specific hour, minute, and second of this day—somewhere along this year-long "cosmic calendar."

Table 22.1 on the worksheet lists a series of global events, along with the number of years ago that each occurred, according to the consensus of scientific researchers. A simple ratio can be used to determine where to place these events on the cosmic calendar, represented by the final column of Table 22.1. Let N represent the number of years in the past that an event occurred, and T the corresponding number of cosmic-calendar days before year's end at midnight, December 31. Then the "count-back" days T can be computed from N according to the relation:

$$T = \frac{N}{4.6 \times 10^9} \times 365. \tag{22.1}$$

For example, suppose an event occurred $N = 100$ million (1×10^8) years ago. According to Equation 22.1, the corresponding number of count-back days T is:

$$T = \frac{1 \times 10^8}{4.6 \times 10^9} \times 365 = 8 \text{ days}.$$

To situate this event in its proper place on the cosmic calendar, count-back 8 days from December 31; the answer is December 23.

For relatively recent events—those that occurred mere thousands or millions of years ago—you will find that the count-back number T comes out to a fraction of a day. In such cases, it is better to convert count-back days into count-back hours, count-back minutes, or even count-back seconds, as follows:

- To convert T count-back days into count-back hours, multiply T by 24 (the number of hours in a day).
- To convert T count-back days into count-back minutes, multiply T by 1440 (the number of minutes in a day).
- To convert T count-back days into count-back seconds, multiply T by 86,400 (the number of seconds in a day).

For example, an event that occurred 1 million years ago corresponds to about 10 P.M. on December 31 of the cosmic calendar. As you might imagine, your birth date would translate into a tiny fraction of a second before the stroke of midnight on the cosmic calendar's last day!

1. Compute the count-back time and the equivalent cosmic-calendar date for each of the events listed in Table 22.1 on the worksheet. Enter your answers into the table.
2. On the worksheet, write a paragraph describing what you learned from this activity about the timeline of the development of life on Earth.

Worksheet, Activity 22: Geological Time in Perspective

Name _____

1.

Table 22.1. Selected Events in the Earth's History

Event	Number of years ago (N)	Count-back time (T) specify days, hours, minutes, or seconds, as appropriate	Cosmic calendar date (and time of day, if appropriate)
Formation of Earth	4.6 billion		
First life on Earth	4 billion		
First plants	2 billion		
Earth's oxygen atmosphere forms	600 million		
First land animals/insects	400 million		
Rise of the dinosaurs	250 million		
Extinction of the dinosaurs	65 million		
First hominids (human-like creatures)	3 million		
Early civilizations	10,000		
Columbus "discovers" America	500		

2. _____
_____
_____
_____
_____
_____
_____
_____

Introduction to Stars and Planets
An activities-based exploration
Alan Hirshfeld

Activity 23

The Comparative Density of Planets

Preview

The solar system's planets fall into two broad categories: Earth-like and Jupiter-like. The physical characteristics and general location of these terrestrial and Jovian types are examined, and their mode of formation considered.

23.1 Average Density

The average density of an object is defined as the object's mass divided by its volume, or its mass per unit volume. While neither stars nor planets are uniformly dense throughout, their average density might give us a rough idea of the chemical composition, concentration, and state of matter within their interior. In general, celestial objects that have a relatively low average density are predominantly gaseous or liquid, while those that have a relatively high average density tend to be in mostly solid form. One caveat to this broad conclusion: whether a substance is gaseous, liquid, or solid depends, not only on its density, but also on the pressure and temperature of its environment. Let's explore the relative densities of the planets to see whether and how they differ and to assess their interior physical state.

The Earth's average density ρ_E (represented traditionally by the Greek letter "rho") is defined by the expression:

$$\rho_E = \frac{M_E}{V_E}, \tag{23.1}$$

where M_E and V_E are the Earth's mass and volume, respectively. The Earth's mass M_E is determined from Kepler's third law, whose input data are the orbital radius and orbital period of any Earth-circling body, such as the Moon or an artificial satellite. (In a previous activity, we used the Earth's own orbital parameters to find the mass of the Sun.) The Earth's volume V_E is computed from the formula for the volume of a sphere, $V_E = (^4/_3)\pi R^3$. Thus, Equation (23.1) can be rewritten:

© IOP Publishing Ltd 2020

$$\rho_E = \frac{M_E}{(4/3)\pi R_E^3}. \tag{23.2}$$

Substituting the numerical values for the Earth's mass and volume into Equation (23.2), our planet's average density ρ_E works out to approximately 5.5 grams per cubic centimeter (g cm^{-3}). By comparison, the density of liquid water is 1.0 g cm^{-3} while that of the common element iron is 7.9 g cm^{-3}. Toward the upper extreme of elemental density is gold, which tips the scales at a hefty 19.3 g cm^{-3}. A basketball-sized volume of gold has a mass of 140 kg, weighing more than 300 pounds! (Gold's high density plays a key role in the cliffhanger ending of the film "The Italian Job" from 1969.)

1. Most of the Earth's surface rocks have densities between 2 and 3 g cm^{-3}. Given that the Earth's overall average density is 5.5 g cm^{-3}, what might you reasonably conclude about the density of matter in our planet's deep interior? Justify your conclusion in a few sentences.

With minor adjustments, Equation (23.2) can be recast to yield the average density ρ_P of other planets:

$$\rho_P = \frac{M_P}{(4/3)\pi R_P^3}, \tag{23.3}$$

where M_P and R_P represent the planet's mass and radius, respectively. However, there's an easier way to accomplish this, in which we express a planet's density as a fraction or multiple of the Earth's density. First we divide the formula for the planet's density, Equation (23.3), by the formula for the Earth's density, Equation (23.2). Next we cancel the terms in common, which yields:

$$\frac{\rho_P}{\rho_E} = \frac{M_P/M_E}{(R_P/R_E)^3}. \tag{23.4}$$

Multiplying both sides of Equation (23.4) by ρ_E gives:

$$\rho_P = \frac{M_{P(E)}}{R_{P(E)}^3} \times \rho_E, \tag{23.5}$$

where $M_{P(E)}$ and $R_{P(E)}$ indicate that the mass and radius of the planet is expressed as a multiple or a fraction of the Earth's mass and radius, respectively.

2. Table 23.1 on the worksheet lists the planets' mass and radius, in Earth units, arranged in order of distance from the Sun. (In these units, the Earth's mass and radius are both equal to 1.00.) Compute the average density of each planet using Equation (23.5) and enter your results in the table.
3. Planets whose average density is close to 1.0 g cm^{-3} are believed to be primarily gaseous or liquid, while those whose average density exceeds about 4.0 g cm^{-3} are believed to be in mostly solid form. Indicate in Table 23.1 which planets are predominantly gaseous/liquid (GL) versus solid (S).

23.2 Terrestrial versus Jovian Planets

4. Astronomers categorize the solar system's planets as either *terrestrial* or *Jovian* depending on whether a planet's physical parameters more closely resemble those of the compact, rocky Earth or the gas-giant Jupiter, respectively. Based on the given planetary mass and radius data, plus your computed average density data in Table 23.1, categorize the planets Mercury through Neptune as either terrestrial (T) or Jovian (J). Enter your results in the table.

5. In Table 23.2, define the terrestrial and Jovian planetary categories in terms of their mass, radius, average density, and proximity to the Sun (whether they are relatively close to the Sun, like the Earth, or else relatively far away).

Astronomers believe that terrestrial planets accreted initially from multitudes of tiny, colliding particles, and later from sub-planetary bodies (*planetesimals*) consisting of rocks and metals. Gases in the inner solar system were swept outward by the solar wind, while ices were vaporized by the flux of solar energy, accounting for their relatively low abundance in terrestrial planets. On the other hand, Jovian planets formed farther from the Sun, where the solar wind was weaker, leaving gases and ices in greater abundance.

6. Compute the average density of the *dwarf planet* Pluto, given its mass and radius data in Table 23.1. Based on your result and the given data, assess whether Pluto belongs in the terrestrial or Jovian category or in neither of these categories.

Pluto, and physically similar bodies in the region of the outer solar system known as the Kuiper Belt, appear to represent a distinct category of icy, moon-size objects orbiting the Sun. While some 2500 Kuiper Belt objects have been observed and characterized to date, tens of thousands more of these small worlds likely exist.

Worksheet, Activity 23: The Comparative Density of Planets

Name _____

1. _____
   _____
   _____

2, 3, 4.

Table 23.1. Physical Data for the Solar System's Planets

Planet	Mass (Earth Units)	Radius (Earth Units)	Density $(g\ cm^{-1})$	Gas/liquid (GL) or Solid (S)	Terrestrial (T) or Jovian (J)
Mercury	0.06	0.38			
Venus	0.82	0.95			
Earth	1.00	1.00	5.5		
Mars	0.11	0.53			
Jupiter	318	11.21			
Saturn	95	9.45			
Uranus	14.5	4.01			
Neptune	17.1	3.88			
Pluto[a]	0.002	0.19			

[a] Dwarf planet.

5.

Table 23.2. Defining Characteristics of Terrestrial and Jovian Planets

	Terrestrial Planets	Jovian Planets
Mass		
Radius		
Average density		
Proximity to the Sun		

6. _____
   _____

Introduction to Stars and Planets
An activities-based exploration
Alan Hirshfeld

Activity 24

Planetary Surface Temperatures

Preview

The surface temperatures of planets and other solar system bodies are calculated from the laws of physics and compared to their measured values. The warming of a planet's surface by the greenhouse effect is considered.

24.1 Introduction

Planets are celestial islands scattered within the vast ocean of outer space. They are a diverse lot, ranging from rocky, airless bodies akin to Mercury in our own solar system to gaseous behemoths that make Jupiter look like a lightweight. The surface temperature on these worlds likewise runs the gamut from well below zero to many hundreds of degrees above zero, depending in large part on the intensity of illumination by their central star. Planets generate only a small amount of thermal energy, if any measurable amount at all; thus, their global surface temperature is the result of a balance between the light energy they receive from their host star and the energy they radiate back into space.

24.2 Earth's Surface Temperature

The nominal surface temperature of the Earth, say, can be predicted through the application of some elementary physics. Let's simplify the problem by ignoring any greenhouse, or warming, effect of the Earth's atmosphere on its global average temperature. Sunlight strikes the ground, which heats up and radiates back a portion of the absorbed solar energy in the form of infrared light. (You might have felt this radiant energy while crossing a paved parking lot during a hot summer's day.) The Earth's surface settles at a more-or-less steady temperature when it reaches a state called *thermal equilibrium*: when the energy the Earth radiates into space balances the energy it absorbs from sunlight.

doi:10.1088/2514-3433/abc249ch24

© IOP Publishing Ltd 2020

In a previous activity, we determined the solar constant S, the luminous power shining directly onto a 1 m^2 area at the Earth: about 1370 watts per square meter (W m^{-2}). To compute the solar power that illuminates, not just 1 m^2, but the Earth's entire Sun-facing hemisphere, we recognize that the energy intercepted by our spherical planet is equivalent to the energy intercepted by a circular disk whose radius is that of the Earth, R_E. Therefore, the total solar power P_{in}, in watts, illuminating the Earth is found by multiplying the solar constant S by the area of this Earth-radius disk:

$$P_{in} = S \times \pi R_E^2. \tag{24.1}$$

If the Earth's surface were perfectly black, virtually all of this incoming solar energy would be absorbed by the ground; on the other hand, if the Earth's surface were entirely ice-covered, most of the incoming energy would be reflected from the ground. The albedo a (pronounced "al-bee-doe") is defined as the fraction of incoming light that is *reflected* from a planet's surface, and ranges from 0.0 (no reflection) to 1.0 (complete reflection). The expression $(1 - a)$ therefore gives the fraction of incoming light *absorbed* by a planet's surface. The Earth's albedo is about 0.30; that is, 30% of solar illumination is reflected by the Earth back into space, while the remaining 70% is absorbed.

Equation (24.1) can now be modified to represent the solar power *absorbed* by Earth's surface (the remainder being reflected):

$$P_{abs} = (1 - a) \times \left(S \times \pi R_E^2 \right). \tag{24.2}$$

Absorbing this solar energy, the Earth's surface heats up and radiates energy according to a formula developed by the 19th-century physicists Josef Stefan and Ludwig Boltzmann:

$$P_{rad} = (\sigma T^4) \times \left(4\pi R_E^2 \right), \tag{24.3}$$

where P_{rad} is the power, in watts, radiated from the Earth's surface; σ (the Greek letter "sigma") is a physical constant whose value is 5.67×10^{-8} in our chosen system of units; T is the Earth's surface temperature, on the Kelvin scale (modern LED light bulbs are characterized by their *color temperature* in Kelvin units); and $4\pi R_E^2$ is the expression for the Earth's global surface area. (The Earth spins once every 24 hr; thus its entire spherical surface is warmed by sunlight and emits energy.)

The condition of thermal equilibrium, described above, implies that the amount of solar energy absorbed by the Earth's surface equals the energy it radiates back into space: $P_{abs} = P_{rad}$. Therefore, we can set Equations (24.2) and (24.3) equal to each other, then cancel out the terms common to both sides, plug in the numerical value of the σ constant, and arrive at the following expression for the Earth's surface temperature:

$$T = 45.82 \times \sqrt[4]{(1 - a) \times S}. \tag{24.4}$$

1. Use Equation (24.4) to compute the Earth's global average surface temperature T in Kelvin units (K). Enter your answer into Table 24.1 on the

worksheet. Note: The 4th root of a quantity can be determined by pressing the square root key on your calculator twice in succession; for example, the 4th root of 16 is 2.

2. Convert your answer for T above into the centigrade system (°C) using the formula °C = K − 273. Enter your answer into Table 24.1 on the worksheet.
3. How does your answer to part 2 above compare to the freezing point of water, 0 °C? From your computed temperature, which is based on proven principles of physics, what conclusion would you draw about the likelihood of the presence of *liquid* water on the Earth's surface?

In actuality, the Earth's global average surface temperature is not nearly as frigid as the number you computed above, but closer to 14 °C. Were it not for the atmospheric greenhouse effect, which traps infrared wavelengths of energy the Earth would otherwise radiate into outer space, our planet might be completely iced over!

24.3 The Surface Temperature of Other Planets

Equation (24.4) can be generalized to other planets in the solar system, once their albedo and "equivalent solar constant" are determined. Table 24.1 on the worksheet lists the albedo of the planets Mercury, Venus, and Mars. (The albedo of Venus is that of its highly reflective blanket of clouds, not its actual surface.) Each planet's equivalent solar constant S_P can be derived using the inverse square law, which states that solar light intensity diminishes with the inverse square of the distance from the Sun. For example, if a planet is located *twice* the Earth's distance from the Sun—that is, at 2 au instead of 1 au—then its equivalent solar constant S_P is *one-fourth* of the value measured at Earth: $^1/_4 \times 1370$, or 343 W m^{-2}. If a planet is located *half* the Earth's distance from the Sun—at 0.5 au—then its equivalent solar constant S_P is *four* times the value measured at Earth: 4×1370, or 5480 W m^{-2}. For parts 4, 5, and 6 below, enter your answers into Table 24.1 on the worksheet.

4. Use the inverse square law to determine the equivalent solar constant S_P at the planets Mercury, Venus, and Mars.
5. Use Equation (24.4) to compute the global average surface temperature T of Mercury, Venus, and Mars in Kelvin units (K).
6. Convert your answers for T above into the centigrade system (°C) using the formula from part 2.
7. For which of the three planets Mercury, Venus, and Mars does the predicted surface temperature approximate the actual surface temperature? Account for any minor discrepancy between the predicted and actual values.
8. For which of the three planets Mercury, Venus, and Mars does the predicted surface temperature differ considerably from the actual surface temperature? Account for the discrepancy between the predicted and actual values.

24.4 The Surface Temperature of a Comet

Equation (24.4) can also be used to predict the surface temperature of non-planetary bodies, such as comets and asteroids. In 1986, the European Space Agency's Giotto spacecraft flew past the nucleus of Halley's comet as the comet neared perihelion, its closest approach to the Sun. Giotto revealed Halley's nucleus to be a pitted, potato-shaped conglomerate of water-ice and silicon- and carbon-rich dust particles, altogether some 10 km across. Much of the surface is coated with pitch-black organic compounds, which absorb fully 96% of incoming sunlight! That is, Halley's albedo *a* is a coal-like 0.04. For parts 9, 10, and 11 below, enter your answers into Table 24.1 on the worksheet.

9. Use the inverse square law to determine the equivalent solar constant at Halley's comet when the comet is at perihelion, about 0.6 au from the Sun.
10. Use Equation (24.4) to compute the predicted surface temperature T, in Kelvin units (K), on the Sun-facing side of the nucleus of Halley's comet.
11. Convert your answer for T above into the centigrade system (°C) using the formula from part 2. Compare your answer for T to the surface temperature measured by the Giotto spacecraft during its 1986 flyby: roughly 77 °C. If your answer differs considerably from the measured value, check your calculations.
12. The freezing point of water at a pressure of 1 Earth-atmosphere is 0 °C. Above this temperature, water-ice melts and turns into a liquid, or in the vacuum of outer space, it transforms (*sublimates*) directly into a gas: water vapor. Given your answer for the surface temperature of Halley's comet, speculate on the long-term effect of solar illumination on the comet, knowing that it swings close to the Sun, on average, every 76 years. What do you predict is Halley's eventual fate?

Worksheet, Activity 24: Planetary Surface Temperatures

Name _____

1, 2.

Table 24.1. Data for Selected Solar System Planets Plus Halley's Comet

Planet/object	Distance (au)	Albedo a	S or S_P (W m^{-2})	$T_{computed}$ (K)	$T_{computed}$ (°C)	T_{actual} (°C)
Earth	1.00	0.30	1370			14
Mercury	0.39	0.07				167
Venus	0.72	0.77[a]				460
Mars	1.52	0.25				−50
Halley's comet	0.6	0.04				

[a] Albedo of atmospheric cloud layer.

3. _____
   _____
   _____

4, 5, 6. See Table 24.1.

7. _____
   _____
   _____

8. _____
   _____
   _____

9, 10, 11. See Table 24.1.

12. _____
    _____

Introduction to Stars and Planets
An activities-based exploration
Alan Hirshfeld

Activity 25

The Habitable Zone

Preview

In their search for extraterrestrial life, astronomers have developed the concept of a habitable zone: a region of space surrounding a star, not too hot, not too cold, where planets might harbor long-lasting bodies of liquid water on their surface. The Earth occupies such a habitable zone around the Sun. But how might our planet's life-sustaining environment change if its central star were swapped out for another brighter or dimmer star?

25.1 Introduction

The Sun is an average star in every sense but one: its third planet, the Earth, harbors the only known example to date of life in the universe. At spectral type G2, the Sun lies roughly midway between the temperature–luminosity extremes of the main sequence in the Hertzsprung–Russell (HR) diagram. (The spectral-type labels for stars are O, B, A, F, G, K, and M, in descending order of surface temperature, with O about 30,000°C–40,000°C and M about 3000°C.) The Sun's lifetime is long enough—around 10 billion years—and its energy outflow steady enough for life to have arisen and prospered on at least one of its surrounding worlds. Our Earth orbits the Sun in a near-circle, maintaining just the right distance to forestall global incineration or freeze-over. At this distance, water exists in liquid form and our life-sustaining atmosphere remains intact despite an unceasing torrent of solar rays and particles.

But what if the situation were different? What if the Sun were swapped out for another star? Stars are a diverse species, ranging widely in mass, size, temperature, luminosity, and longevity. Moving upward along the HR diagram's main sequence, we find hotter, more brilliant emitters than the Sun: a star of spectral type A0, such as Vega in the constellation Lyra, blazes at more than 10,000 °C on its surface and puts out approximately 80 times the Sun's energy every second. Moving downward along the main sequence, we encounter progressively dimmer, redder stars: a star of spectral type M0, among the most common in the Galaxy, has a surface temperature under 4000 °C and emits a measly 0.06 times the Sun's luminosity.

25.2 Replacing the Sun

Imagine the Sun booted off to some remote corner of the universe and replaced by another star. How might Earth's habitability change if, instead of being nourished by the favorable rays of the Sun, our planet were exposed to the harsh intensity of an A0 star or the dusky glimmer of an M0 star? In previous activities, we computed the intensity of sunlight arriving at the Earth, as well as the physics that determines the Earth's surface temperature. Together these lines of exploration allow us a scientifically informed peek at what the "solar" system might be like if a different star presided over the planets.

The solar constant S is the luminous power shining on a 1 m^2 area located at the Earth, 1 au from the Sun. Its value can be computed from the formula:

$$S = \frac{L}{4\pi r_E^2},\tag{25.1}$$

where L is the Sun's luminosity and r_E is the Earth's distance from the Sun. This equation is one example of an inverse square law (Newton's universal law of gravitation is another); here, the intensity of sunlight diminishes with the square of the distance, the result of the Sun's emission spreading out over a progressively larger volume of space. If we were to simply replace the Sun with another star and leave the Earth where it is, at the distance r_E, Equation 25.1 indicates that the new "solar constant" at the Earth depends only on the luminosity of the replacement star. If the replacement star's luminosity is, say, twice the Sun's luminosity, then the adjusted solar constant S would be double its current value. (For simplicity, we will ignore the differing wavelength distributions of solar emission versus those of other stars.)

The revised solar constant will, in turn, factor into the determination of the Earth's average surface temperature. As we have already seen in a previous activity, the key formula arises from the balance between the incoming light energy from the central star and the outgoing energy radiated by the Earth into space. This formula is expressed as follows:

$$T = 45.82 \times \sqrt[4]{(1 - a) \times S},\tag{25.2}$$

where T is the Earth's surface temperature in Kelvin units (subtract 273 to convert to the centigrade scale), a is the Earth's albedo (0.30), and S is the solar constant in W m^{-2}.

1. Suppose the Sun were replaced by a main sequence star of spectral type A0, whose luminosity is 80 times that of the Sun. (a) Considering Equation 25.1, how would an eighty-fold increase in luminosity affect the value of the solar constant S measured at the Earth? (b) Compute the revised solar constant S_{A0}, in watts per square meter (W m^{-2}), if an A0 star were situated at the center of the solar system. The current solar constant S has a value of 1370 W m^{-2}.

2. (a) Substitute the revised solar constant S_{A0} from part 1(b) into Equation 25.2 and compute the Earth's predicted surface temperature T, in both Kelvin and centigrade units. Assume that the Earth's albedo remains unchanged, at 0.30. (b) Write a statement evaluating the likelihood of the long-term presence of liquid water on the Earth's surface given the predicted surface temperature from part 2(a). How might this altered temperature affect Earth's habitability?

3. Suppose the Sun were now replaced by a main sequence star of spectral type M0, whose luminosity is 0.06 times that of the Sun. (a) Considering Equation 25.1, how would this decrease in luminosity affect the value of the solar constant S measured at the Earth? (b) Compute the revised solar constant S_{M0}, in W m^{-2}, if an M0 star were situated at the center of the solar system.

4. (a) Substitute the revised solar constant S_{M0} from part 3(b) into Equation 25.2 and compute the Earth's predicted surface temperature T, in both Kelvin and centigrade units. Again, assume that the Earth's albedo remains unchanged, at 0.30. (b) Write a statement evaluating the likelihood of the long-term presence of liquid water on the Earth's surface given the predicted surface temperature from part 4(a). How might this altered temperature affect the Earth's habitability?

25.3 Relocating the Earth

Now let's evaluate Earth's habitability from a somewhat different perspective: how far away from its substitute central star would Earth have to be placed in order to maintain its current average surface temperature, that is, to remain habitable? The solution to this question lies in Equation 25.1 above. In replacing the Sun with another star, we alter the numerical value of L on the right side of the equation. Yet at the same time, we require that the numerical value of the solar constant S, our nominal habitability index, stays the same. The only way that can happen is if we alter the value of $r_E{}^2$ in the denominator such that it "cancels out" the change in the numerator L.

For example, if the L is doubled, then the quantity $r_E{}^2$ must likewise be doubled, in order that S remain the same. But to double the squared term $r_E{}^2$, we would multiply r_E itself by the *square root* of 2. Were we to replace the Sun with a star, say, 36 times as luminous, yet receive the same amount of solar energy as we do now, we would have to relocate the Earth from its current distance of from 1 au to a new distance of 6 au.

5. Suppose the Sun were replaced by a main sequence star of spectral type A0, whose luminosity is 80 times that of the Sun. (a) Considering Equation 25.1, how far from the A0 star, in au, must the Earth be situated such that the current solar constant remains unchanged? (b) Compare your answer to the list of planetary distances in Table 25.1. Where would the Earth's revised distance place it relative to the locations of the planets?

6. Suppose the Sun were replaced by a main sequence star of spectral type M0, whose luminosity is a mere 0.06 times that of the Sun. To maintain the

current solar constant, the Earth must be moved *closer* to the central star. (a) Considering Equation 25.1, how far from the M0 star, in au, must the Earth be situated to fulfill this condition? (b) Compare your answer to the list of planetary distances in Table 25.1. Where would the Earth's revised distance place it relative to the locations of the planets?

Planetary surface temperature is only a rough indicator of a planet's habitability, here narrowly defined by the temperature range that permits long-term retention of bodies of liquid water. Other factors figure into the habitability of a world, such as planetary size, presence of an atmosphere, and the warming influence of the greenhouse effect. (Too much greenhouse effect renders a planet inhabitable, as in the case of Venus, whose surface broils at 460°C, almost 900°F.) Nevertheless, our simple physics model supports the logical conclusion that the so-called habitable zone lies farther from a hot, luminous star than from a cool, dim star.

Physics provides only one perspective on the issue of planetary habitability; the problem must be tackled from the biological side as well. How does life arise in the first place? Given the proper conditions, is the establishment of life rare or commonplace, if not unique to our world alone? And, as Earthlings, are our notions about the possible variety of life forms too constricted? Even here on our home planet, there exist organisms—*extremophiles*—that survive in lightless, highly acidic, or ultra-hot or ultra-cold environments. Until we observe some indisputable evidence of living creatures off-Earth or brew organisms from scratch in the lab, the subject of extraterrestrial life will remain a speculative endeavor.

But in the meantime, advancing technology has given astronomers the observational tools to escape the realm of the hypothetical and embark on an all-out search for *real* exoplanets. As we will learn, subtle wobbles in stars' movements or periodic winks in their light output have revealed the presence of a multitude of worlds beyond our solar system.

Table 25.1. Average Distance of the Solar System's Planets from the Sun.

Planet	Distance (au)
Mercury	0.39
Venus	0.72
Earth	1.00
Mars	1.52
Jupiter	5.20
Saturn	9.58
Uranus	19.2
Neptune	30.1

Worksheet, Activity 25: The Habitable Zone

Name _____

1. (a) _____

   (b) $S_{A0} =$ _____ W m^{-2}

2. (a) $T =$ _____ K = _____ °C

   (b) _____

3. (a) _____

   (b) $S_{M0} =$ _____ W m^{-2}

4. (a) $T =$ _____ K = _____ °C

   (b) _____
   _____

5. (a) Earth's distance from A0 star = _____ au

   (b) _____

6. (a) Earth's distance from M0 star = _____ au

   (b) _____

Introduction to Stars and Planets
An activities-based exploration
Alan Hirshfeld

Activity 26

The Search for Exoplanets: Doppler Method

Preview

By searching for periodic shifts in the wavelengths of a star's spectral lines, astronomers are able to detect the presence of planets around stars outside our solar system: *exoplanets*. Measurement of these wavelength shifts, which arise from the Doppler effect, reveal orbital parameters of the star–exoplanet pair as well as the exoplanet's mass.

26.1 The Doppler Effect

In 1842, Austrian mathematician Christian Doppler proposed that the perceived frequency, or pitch, of sound waves changes depending on the relative motion of the sound source and the listener: the pitch rises when source and listener are approaching one another, and falls when they are moving apart. Doppler suggested that the effect holds for all types of waves, including light. Wavelengths of light from an approaching star would be shortened, or *blueshifted*, compared to their nominal values for a stationary star, whereas wavelengths of light from a receding star would be lengthened, or *redshifted*. In the case of a binary star system, viewed along their orbital plane, the Doppler shift of each member's light would be cyclic, oscillating between blueshift and redshift, as the stars circle each other.

In actuality, the Doppler effect produces no discernible change in a star's overall color; whatever light energy is shifted out of a particular segment of the color palette is replaced by other light energy shifted into this segment. However, the Doppler effect *can* produce a measurable change in the wavelength of a star's spectral lines. Observing such spectral-line shifts, astronomers are able to determine whether the line-of-sight, or radial, component of a star's velocity trends toward or away from the Earth. And if those spectral-line shifts are periodic, they reveal the underlying presence of a binary star, including those too faint to be imaged directly through a telescope.

doi:10.1088/2514-3433/abc249ch26

© IOP Publishing Ltd 2020

Today's astronomical spectrometers are so refined that even a slight variation in a star's radial velocity is detectable in the Doppler shift of its spectral lines. This has given astronomers the ability to spot the telltale gravitational tug that a planet exerts on its host star as they move around one another. Advancements in telescope and detector technology have led to a large and growing list of *exoplanets* discovered by the Doppler method.

26.2 Center of Mass

A star is commonly depicted as the stationary center of its planetary system, with planets circling that fixed locus in space. In reality, each planet exerts a gravitational pull that displaces the star, to a greater or lesser degree, from that central location. Thus, the star and the planet mutually orbit a point called the *center of mass* that lies on a line between the pair, as pictured in Figure 26.1. The star executes a tiny orbit of its own, whose overall size and shape are determined by the masses and changing orbital positions of its planetary retinue. Astronomers are able to characterize these planets by analyzing the star's movement, as described below.

Imagine a pair of celestial bodies in orbit around one another. If the bodies are identical, their center of mass will lie midway between them; if their masses are unequal, the center of mass will lie closer to the more massive body, as is the case in Figure 26.1. The placement of a star and a planet relative to their center of mass is expressed mathematically by a simple formula:

$$M_{\text{star}} \times r_{\text{star}} = M_{\text{planet}} \times r_{\text{planet}}, \qquad (26.1)$$

where M_{star} and M_{planet} are the masses of the star and planet, respectively, and r_{star} and r_{planet} are their respective distances from the center of mass. (M_{star} and M_{planet}

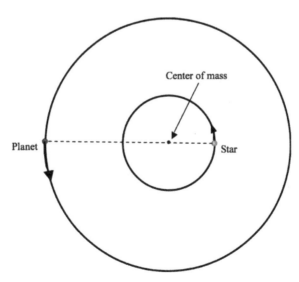

Figure 26.1. Center of mass of a star–planet system, with arrows indicating their relative velocities.

are expressed in the same units—solar masses, Earth masses, kilograms, etc.—as are r_{star} and r_{planet}). Equation (26.1) can be solved for r_{star} to yield:

$$r_{star} = \left(\frac{M_{planet}}{M_{star}} \right) \times r_{planet}. \tag{26.2}$$

1. Consider the mass ratio within parentheses in Equation (26.2). Since the mass of a planet is typically a tiny fraction of a star's mass, what does the formula predict about the star's distance from the center of mass r_{star} compared to the planet's distance from the center of mass r_{planet}?

Having derived the star's distance from the center of mass in Equation (26.2), we next ask how fast the star circulates around the center of mass. Let's assume that the star's orbit is circular. Then the velocity of the star v_{star} around the center of mass is equal to (i) the distance traveled by the star—here, the circumference of its orbit $2\pi r_{star}$—divided by (ii) the time it takes the star to cover this distance —its orbital period P:

$$v_{star} = \frac{2\pi r_{star}}{P}. \tag{26.3}$$

2. Note that the star and the planet always lie directly opposite each other as they move about their common center of mass. What does this imply about the orbital period of the star compared to that of the planet?

26.3 Detecting the Solar System from Afar

Let's apply the basic concepts described above to the following question: using present-day technology, could an astronomer on a distant world detect the presence of, say, the planet Jupiter from subtle variations in the Sun's radial velocity? Here's the essential data: the Sun's mass is 333,000 Earths; Jupiter's mass is 318 Earths, its orbital period 11.9 yr (3.75×10^8 s), and its orbital radius 5.2 au (7.8×10^8 km).

3. (a) Use Equation (26.2) to compute the distance r_{star} of the Sun's center from the Sun–Jupiter center of mass, in kilometers. (b) Write a statement that describes the location of the Sun–Jupiter center of mass relative to the location of the Sun's surface. (The Sun's radius is approximately 700,000 km.)
4. Use Equation (26.3) to compute the Sun's orbital velocity v_{star} around the Sun–Jupiter center of mass, in kilometers per second. Convert your answer to meters per second. (There are 1000 m in a kilometer.)
5. (a) Use Equation (26.2) to compute the distance r_{star} of the Sun's center from the Sun–Earth center of mass, in kilometers. (The Earth is 1 au, or 1.5×10^8 km, away from the Sun.) (b) Write a statement that describes the location of the Sun–Earth center of mass relative to the Sun's geometric center.

6. Use Equation (26.3) to compute the Sun's orbital velocity v_{star} around the Sun–Earth center of mass, in kilometers per second. Convert your answer to meters per second.
7. The High Accuracy Radial-Velocity Planet Searcher (HARPS), coupled to the European Southern Observatory's 3.6 m telescope at La Silla, Chile, can measure stellar velocity deviations as small as 0.5 m s^{-1}. From your results above, could an extraterrestrial astronomer using HARPS-equivalent technology detect the Sun's orbital velocity variations (a) due to the presence of Jupiter? (b) due to the presence of the Earth?
8. Using the Doppler method, would it be easier to detect (i) an Earthlike planet orbiting a star like the Sun or (ii) an Earthlike planet orbiting a "red dwarf" star only a few tenths of the Sun's mass? Explain your reasoning.

26.4 Weighing an Exoplanet

In 1995, astronomers observed that the radial velocity of the star 51 Pegasi varies periodically by up to 56 m s^{-1} (0.056 km s^{-1}) over a time span of 4.23 days (3.65 × 10^5 s). Evidently, 51 Pegasi is alternately swinging toward, then away from the Earth, as if circling an unseen body, as depicted in Figure 26.2. The degree and regularity of the velocity oscillation have led astronomers to conclude that the star is paired with a planet too faint to be imaged through a telescope. Using the principles described above, we can deduce the mass of 51 Pegasi's companion world. The first step is to recast the center-of-mass formula, Equation (26.1), as follows:

$$M_{planet} = \left(\frac{r_{star}}{r_{planet}} \right) \times M_{star}. \tag{26.4}$$

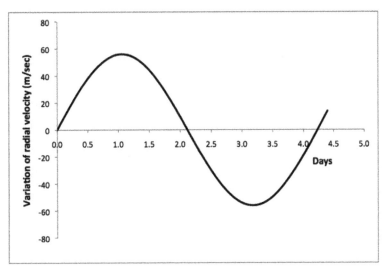

Figure 26.2. Graphical representation of the periodic radial-velocity variations of the star 51 Pegasi.

9. Even before plugging in the data, we can rough-check the validity of the formula. Consider the ratio of the orbital radii within parentheses in Equation (26.4) in light of what you have learned about the center of mass of a star–planet system. Does Equation (26.4) correctly predict that a planet's mass is typically a tiny fraction of its host star's mass? Explain your reasoning.

According to Equation (26.4), in order to compute the mass of 51 Pegasi's planet M_{planet}, we must quantify the three variables to the right of the equal sign: (i) the mass of the star M_{star}; (ii) the radius of the star's orbit r_{star}; and (iii) the radius of the planet's orbit r_{planet}.

• The star 51 Pegasi is very much like our Sun in its observable properties, from which we conclude that M_{star} must be very close to 1 solar mass (333,000 Earths).
• We obtain the radius of the star's orbit r_{star} from a simple rearrangement of Equation (26.3):

$$r_{star}(\text{km}) = \frac{P v_{star}}{2\pi}. \tag{26.5}$$

Here, the orbital period P is expressed in seconds and the orbital velocity v_{star} in kilometers per second; the resultant orbital radius r_{star} comes out in kilometers.

• As to the radius of the planet's orbit r_{planet}, we make use of Kepler's third law of orbital motion, assuming that the planet's mass is negligible compared to that of the star:

$$r_{planet}^{3} = \left(\frac{G M_{star}}{4\pi^2} \right) \times P^2. \tag{26.6}$$

Solving Equation (26.6) for r_{planet}—that is, taking the cube root of both sides—and substituting the numerical value of the constant G yields the simplified expression:

$$r_{planet}(\text{km}) = 21.6 \times \sqrt[3]{M_{star} P^2}. \tag{26.7}$$

In this formula, r_{planet} comes out in units of kilometers, given the star mass M_{star} in Earth-masses and period P in seconds.

10. Compute the radius of 51 Pegasi's orbit r_{star}, in kilometers, by substituting into Equation (26.5) the observed values of P, in seconds, and v_{star}, in kilometers per second. Note: We have assumed that the orbit of 51 Pegasi is seen edge-on, so that the peak observed radial-velocity shift coincides with the star's true orbital velocity, that is, the star is headed straight toward or away from us at these times; if the orbit is inclined to our view, the orbital velocity is larger than the measured radial-velocity shift of 0.056 km s^{-1}.

11. (a) Compute the radius of the planet's orbit r_{planet}, in kilometers, by substituting into Equation (26.7) the observed values of M_{star}, in Earth-masses, and P, in seconds. (Be sure to take the *cube* root, not the square

Figure 26.3. Exoplanet 51 Pegasi b, an example of a *hot Jupiter*: a gas-giant planet that orbits close to its host star. (Credit: NASA/JPL-Caltech.)

root, of the expression.) (b) To put the size of this planetary orbit into context, write a statement comparing it to the radius of Mercury's orbit around the Sun: 0.39 au, or 5.85×10^7 km.

12. (a) Use the values of M_{star}, r_{star} and r_{planet} found above to solve Equation (26.4) for the mass of the planet circling 51 Pegasi, first expressed in Earth-masses, then in Jupiter-masses. (Jupiter's mass is 318 times that of the Earth.) (b) Is your answer consistent with astronomers' consensus that the mass of 51 Pegasi's planet is about 0.44 Jupiter-masses? If not, check your work.

The planetary companion of 51 Pegasi, designated 51 Pegasi b, became the prototype of a category of celestial bodies named *hot Jupiters*: large, gaseous planets that orbit unexpectedly close to their host star, as in the artist's depiction, Figure 26.3. Still unresolved are whether these hot Jupiters formed where they are or migrated into place, and how they retain their gases in the face of an intense barrage of photons and blistering stellar winds.

Worksheet, Activity 26: The Search for Exoplanets: Doppler Method

Name _____

1. _____
_____

2. _____

3. (a) r_{star} = _____ km

(b) _____
_____

4. v_{star} = _____ km s^{-1} = _____ m s^{-1}

5. (a) r_{star} = _____ km

(b) _____
_____

6. v_{star} = _____ km s^{-1} = _____ m s^{-1}

7. (a) _____

(b) _____

8. _____
_____
_____

9. _____
   _____

10. r_{star} = _____ km

11. (a) r_{planet} = _____ km

    (b) _____
        _____

12. (a) M_{planet} = _____ Earth-masses = _____ Jupiter-masses

    (b) _____

Introduction to Stars and Planets
An activities-based exploration
Alan Hirshfeld

Activity 27

The Search for Exoplanets: Transit Method (I)

Preview

Another way to discover exoplanets is to look for a slight, periodic dimming of a star's brightness, which in some cases might arise from the passage, or *transit*, of a planet across the face of the star. The progression of a transit is often represented in a *light curve*, a graph of the star's brightness before, during, and after the transit. NASA's Kepler satellite, launched in 2009, was designed to search for such exoplanet events among a selection of Sun-like stars. Kepler's sensitivity is examined by checking whether the spacecraft could detect an Earth-transit across the Sun's disk from a location far outside the solar system.

27.1 Introduction

Having seen how exoplanets are detected by the Doppler shift of stellar spectral lines, we now introduce another widely adopted method of uncovering these dim, faraway bodies. Astronomers have long supposed that when a planet crosses the face of its host star—a *planetary transit*—the star's brightness will undergo a minute dip in its brightness. This brightness change will occur on a regular basis, in synchrony with the planet's orbital period around the star. However, it was not until relatively recently that technology improved to the point that the tiny brightness change accompanying a transit became detectable.

In 2009, NASA launched the Kepler satellite, which was dedicated to the discovery of Earth-like exoplanets around Sun-like stars. Kepler's successor, the Transiting Exoplanet Survey Satellite (TESS), was lofted into orbit in 2018. Its mission: ongoing all-sky inspections of 200,000 bright stars near the Sun in search of transiting exoplanets. Utilizing these powerful instruments, the study of exoplanets has evolved from a subject largely of speculation into a full-fledged, data-driven branch of astronomy.

27.2 Planetary Transit Basics

Figure 27.1 depicts the progress, in one-hour increments, of a hypothetical exoplanet before, during, and after it transits along the equator of its host star, as viewed by an Earth-based observer. (The figure is for illustration only; no terrestrial telescope is yet powerful enough to portray extrasolar stars and planets as measurable disks.)

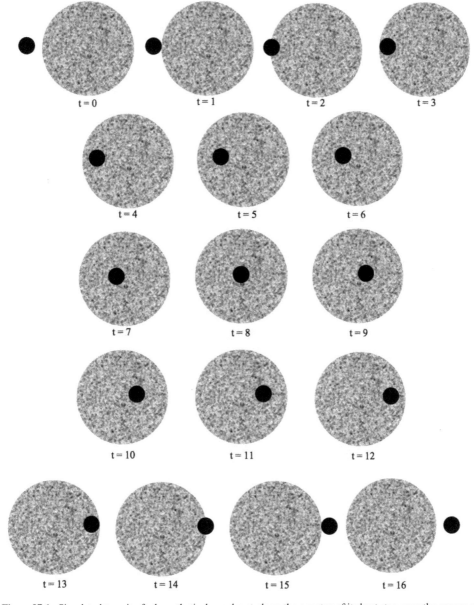

Figure 27.1. Simulated transit of a hypothetical exoplanet along the equator of its host star, over the course of 16 hours.

Let's make the reasonable assumption that the star's measured brightness is determined by the area of its observable disk, the two-dimensional projection of the star's three-dimensional, curved surface.

During a planetary transit, a portion of the star's disk—hence, a portion of its light—is eclipsed by the planet, and the star will appear slightly dimmer than usual. The fractional decrease f in the star's perceived brightness depends on how much of its light is blocked by the planet, that is, it depends on the planet's disk area relative to that of the star. Since the area of a circle is given by the expression πR^2, we can write:

$$f = \pi R_{planet}^2 / \pi R_{star}^2 = (R_{planet}/R_{star})^2. \qquad (27.1)$$

For the purpose of measurement, the ratio of the radii in Equation (27.1) can just as well be expressed as the ratio of the diameters; either way, we arrive at the same answer for f. (Multiplying both the numerator and denominator of a fraction by the same number leaves the fraction unchanged.)

1. In Figure 27.1, align a slip of paper alongside a diameter of the planet (any of the small, darkened circles). With your pencil, mark off the length of the planet's diameter. This pair of marks will serve as our measurement unit: 1 planet-diameter. Move your marked paper over to any of the star disks in Figure 27.1 and, by successively sliding the paper from one edge of the star's disk to the other, measure how many planet-diameters fit across the star's diameter. Your answer might include a fraction of a planet-diameter. (If you prefer, you can instead use a metric ruler to make these measurements in millimeters.)

2. (a) From the previous part's measurements, compute the value of f in Equation (27.1). For example, if the star's diameter is equal to 10 planet-diameters, then the fraction f equals $(1/10)^2$, or $1/100$. The star's light would diminish by $1/100$, or 1%, of its usual value; its perceived brightness during transit would be 99% of that outside of transit. (b) Compute $1.00 - f$, the perceived brightness of the star during transit, where the number 1.00 represents the star's brightness outside of transit.

3. Using the axes of Figure 27.2 on the worksheet, plot the brightness data points of the planetary transit depicted in Figure 27.1, from $t = 0$ to $t = 16$ hr. The brightness of the star outside of transit is indicated on the vertical axis by the number 1.00. Note that there are two instances, at $t = 2$ and $t = 14$ hr, when only part of the planet obscures the star; you will have to compute or estimate the star's perceived brightness at these times. Draw a smooth curve connecting all the data points.

4. (a) Using a dashed line, indicate in Figure 27.2 how the shape of the light curve would change if the planet were smaller than depicted in Figure 27.1. Label your modified light curve "smaller planet."
   (b) Again using a dashed line, indicate on Figure 27.2 how the shape of the light curve would change if the planet transited the star somewhat

off-center, either above or below the equator. Label the modified light curve "off-center."

(c) Again using a dashed line, indicate on Figure 27.2 how the shape of the light curve would change if the planet transited along the bottom edge of the star, that is, if the planet's disk was never fully immersed within the stellar disk. Label the modified light curve "edge transit."

The planetary transit light curves you drew in Figure 27.2 are a simplification of actual observed light curves for another reason (among many). We had assumed that the brightness across the face of the star is uniform. In fact, a glance at a photograph of the Sun shows that this is not the case. The central portion of the Sun's disk is noticeably brighter than areas near its periphery, an effect known as *limb darkening*. The explanation is simple: light from the central region of the solar disk emerges from deeper, hotter, more luminous interior layers than the light near the limb, which is sourced from shallower, cooler, less luminous strata in the solar atmosphere. The same holds true for any star. So as a planet transits a star, its disk masks regions of different brightness, from one limb to the center and on to the opposite limb.

5. Describe the effect the phenomenon of limb darkening would have on the shape of idealized light curves you drew in Figure 27.2. Include a sketch of the revised light curve. Hint: This is a tough one! An analogy might help. Imagine a pair of disks, each hiding an identical area of a light bulb, one a 40 W bulb and the other a much brighter 100 W bulb. In which case would more light be blocked? Apply the same logic to the limb darkening phenomenon and infer the consequent effect on the idealized light curve of a planetary transit.

27.3 Detecting the Solar System from Afar (Redux)

Now that we have explored the general form of a planetary transit light curve, we ask whether planets in our own solar system are detectable from afar by this method, assuming our current level of technology. It is straightforward to calculate the *depth* and the *duration* of the light curve representing a transit of, say, the Earth across the face of the Sun, as seen by a distant observer. The first of these factors—depth of the light curve—imposes a requirement on the sensitivity of the astronomer's observing equipment, specifically, the brightness-measuring device must be sensitive enough to record a very small change in the Sun's light. The second factor—duration of the light curve—sets the frequency with which the astronomer must conduct the solar monitoring: too seldom and an Earth-transit might be missed.

27.3.1 Depth of an Earth-transit Light Curve

6. (a) According to Equation (27.1), when the Earth is silhouetted against the solar disk, the fractional decrease of the Sun's perceived brightness is given by the square of the ratio of Earth's radius to the Sun's radius. Use

Equation (27.1) to compute the fractional decrease f of the Sun's brightness during an Earth-transit. The Sun's radius R_{Sun} is 696,000 km and the Earth's radius R_{Earth} is 6370 km. (b) Could such a decrease in brightness be effectively shown on the axes of Figure 27.2? Explain your answer.

Without a doubt, the solar brightness dip that you computed in part 6 is incredibly small. Yet NASA's Kepler spacecraft was designed to detect such transits of Earth-like planets around Sun-like stars with brightnesses down to apparent magnitude $m = 12$. (Review Activity 13 for the essentials of the astronomical magnitude system of brightness.) Therefore, we ask: how far outside our solar system could a spacecraft with Kepler's sensitivity detect an Earth-transit across the Sun? Expressed differently, how far from its present position, 1 au from the Earth, would the Sun have to be relocated such that it will have faded to Kepler's working magnitude limit of $m = 12$? The key to the solution is the inverse square law, the mathematical link between distance and light intensity.

7. (a) Compute the magnitude difference between the Sun's current magnitude $m_1 = -26.7$ and the target value of $m_2 = 12$.
   (b) Convert this magnitude difference to a brightness ratio, recalling that the brightness ratio is approximated by the expression $2.5^{m_2-m_1}$.
   (c) Light intensity diminishes with the inverse square of the distance; reversing this principle, distance is derived by taking the *square root* of the brightness ratio. Compute the distance, in au, at which the Sun would appear from the Earth as a star of magnitude 12.
   (d) Convert your answer to part 7(c) from au into light-years, given that there are approximately 63,000 au in one light-year. Your answer is the maximum range at which a Kepler-type spacecraft could detect Earth-transits across the face of the Sun.

27.3.2 Duration of an Earth-transit Light Curve

Figure 27.3 shows the Earth—the small circle of radius R_{Earth}—at the start of a transit, which here proceeds from left to right along the Sun's equator. Notice that the point labeled **A** on Earth's leading edge coincides with the leftward limb of the Sun, the larger circle of radius R_{Sun}. The transit continues until the Earth no longer blocks sunlight from reaching the observer; this situation is depicted by the small circle on the right. Notice that the point labeled **A** evidently passed the Sun's rightward limb a while ago, and is situated one Earth-diameter from that limb.

Overall, the transit's duration T equals the time it takes point **A** to travel across the Sun's disk—a distance of $2R_{Sun}$—plus an additional distance equal to the Earth's disk-diameter—$2R_{Earth}$—for a total distance of $2R_{Sun} + 2R_{Earth}$, or $2(R_{Sun} + R_{Earth})$. How quickly this passage occurs depends on the Earth's orbital velocity v around the Sun; hence, the transit duration T is deduced from the conventional formula relating distance, velocity, and time:

$$T = \frac{\text{distance}}{\text{velocity}} = \frac{2(R_{\text{Sun}} + R_{\text{Earth}})}{v}. \tag{27.2}$$

Here, R_{Sun} and R_{Earth} are both known quantities: 696,000 km and 6370 km, respectively. The Earth's orbital velocity v, in the denominator of Equation (27.2), is found by a reapplication of the distance–velocity–time formula, now involving the circumference of the Earth's orbit (distance) and the Earth's orbital period (time):

$$v = \frac{\text{distance}}{\text{time}} = \frac{2\pi r_{\text{Earth}}}{P}, \tag{27.3}$$

where the radius of the Earth's orbit $r_{\text{Earth}} = 1.5 \times 10^8$ km, and the Earth's orbital period $P = 3.15 \times 10^7$ s.

8. Use Equation (27.3), along with the relevant data above, to compute the Earth's orbital velocity v, in kilometers per second.
9. Use Equation (27.2), along with the relevant data above, to compute the duration T of the Earth's transit across the solar disk, in seconds. Convert your answer to hours.

Of course, an extraterrestrial astronomer monitoring the Sun for a planetary transit would initially be unaware of the Earth's existence. Not until the observation of a transit—in fact, several consecutive transits—would this astronomer justifiably assert that the Sun harbored a world that passed across its face once a year. This

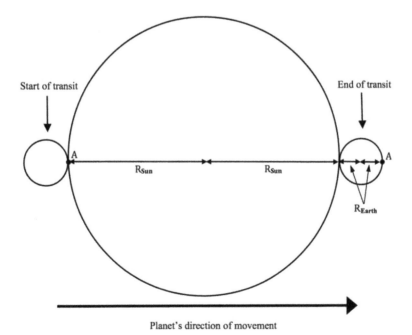

Figure 27.3. Geometry of an Earth-transit along the Sun's equator, as seen by an off-Earth observer.

observer would have to be diligent in keeping track of the solar brightness, so as not to miss the infrequent light-blip that accompanies an Earth-transit. So too, with our own search for exoplanet transits around other stars.

10. Considering your answer to part 9, how often—once a year, once a month, once a day, etc.—do you think the extraterrestrial astronomer should measure the Sun's brightness in order to have a reasonable chance of observing an Earth-transit across the solar disk?

One additional factor that shapes an effective search for exoplanets is the requirement that the plane of the exoplanet's orbit around its host star coincide or nearly coincide with the line of sight from Earth. If the orbital plane is tipped even a few degrees from our viewpoint, the exoplanet will pass either above or below its star and no transit will be seen. Astronomers estimate that, among the general star population, the odds of such an alignment are only about 1 in 200. Not only must exoplanet hunters observe stars frequently and with ultra-fine precision, they must do so for a huge number of stars. The observing list of the Kepler spacecraft runs to more than 150,000 stars, while that of NASA's next-generation Transiting Exoplanet Survey Satellite (TESS) tops 200,000!

Worksheet, Activity 27: The Search for Exoplanets: Transit Method (I)

Name _____

1. _____ planet-diameters (or mm)

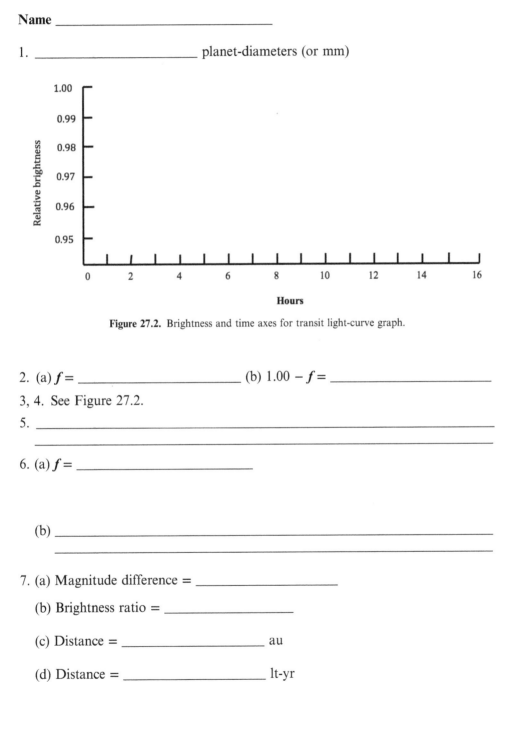

Figure 27.2. Brightness and time axes for transit light-curve graph.

2. (a) f = _____ (b) $1.00 - f$ = _____

3, 4. See Figure 27.2.

5. _____
   _____

6. (a) f = _____

   (b) _____
   _____

7. (a) Magnitude difference = _____

   (b) Brightness ratio = _____

   (c) Distance = _____ au

   (d) Distance = _____ lt-yr

8. $v =$ _____ km s^{-1}

9. $T =$ _____ s $=$ _____ hr

10. Frequency of measurement $=$ _____

Introduction to Stars and Planets
An activities-based exploration
Alan Hirshfeld

Activity 28

The Search for Exoplanets: Transit Method (II)

Preview

During its working lifetime, the Kepler spacecraft discovered more than 2600 exoplanets by the transit method. One such example, called Kepler 21b, is studied here. From the form, depth, and duration of the host star's light curve plus supplementary data, we can deduce Kepler 21b's radius, orbital period, distance from the star, mass, average density, and surface temperature.

28.1 Case Study: The Transit of a Real Exoplanet

In 2011, Kepler mission scientists confirmed the existence of an exoplanet circling the star HD 179070, a relatively bright and well-studied star in the constellation of Lyra. (The designation HD refers to the voluminous *Henry Draper Catalogue* of stellar spectral types, completed in the early 20th century.) An intensive series of space-based and ground-based observations have pinned down the properties of this exoplanet, referred to as Kepler 21b. HD 179070 has a spectral type F6 IV, which indicates that it is somewhat larger, more luminous, higher-mass, and hotter on its surface than the Sun, as summarized in Table 28.1. Kepler 21b transits close to the star's equator once every 2.7858 days (2.41×10^5 s), which we take as the planet's orbital period P.

Figure 28.1 depicts the transit light curve, averaged from hundreds of individual measurements by the Kepler spacecraft's optical detector. (Note the curve's rounded form, the result of the stellar limb-darkening effect described in the previous activity.) High-precision radial-velocity observations have also been completed, revealing the subtle movement of the host star around the system's center of mass. From this wealth of data, we can compute a variety of Kepler 21b's physical properties.

doi:10.1088/2514-3433/abc249ch28

© IOP Publishing Ltd 2020

Table 28.1. Data for Exoplanet Kepler 21b's Host Star

Catalog designation	HD 179070
Apparent magnitude	8.25
Spectral type	F6 IV
Mass	$1.4\ M_{\text{Sun}} = 467{,}000\ M_{\text{Earth}}$
Radius	$1.9\ R_{\text{Sun}} \cong 207\ R_{\text{Earth}} = 1.32 \times 10^{6}\ \text{km}$
Surface temperature	6000°C
Luminosity	$5.2\ L_{\text{Sun}}$
Orbital velocity[a]	$2\ \text{m s}^{-1}$ (approx.)

[a] Note. Velocity around the star–planet center of mass.

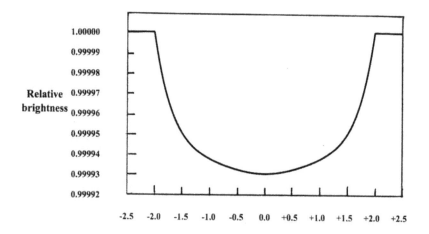

Hours before (-) or after (+) minimum light

Figure 28.1. Averaged light curve of the transit of Kepler 21b across its host star, depicting the relative brightness of the star over a several hour period.

28.2 Radius of Kepler 21b

The planet's radius is obtained from the maximum depth of its transit light curve, as discussed in the previous activity. To this end, we solve Equation (27.1) from that activity for R_{planet}:

$$R_{\text{planet}} = R_{\text{star}} \times \sqrt{f}, \tag{28.1}$$

where f is the fractional reduction of the host star's brightness at mid-transit. R_{planet} takes on the same measurement unit as R_{star}, whether kilometers, Earth-radii, solar-radii, etc.

1. Estimate f at mid-transit from the light curve in Figure 28.1. Remember, brightness is expressed as a fraction of the star's out-of-transit brightness, which is set at 1.00.

2. (a) Compute the radius R_{planet} of Kepler 21b, in units of Earth-radii, from Equation (28.1). Enter your answer in the appropriate row of Table 28.2 on the worksheet. (b) In terms of its radius, is Kepler 21b more like the Earth or more like Jupiter, whose radius is about 11 Earth-radii?

28.3 Orbital Radius of Kepler 21b

In a previous activity, we derived the orbital radius of the 51 Pegasi exoplanet using a simplified form of Kepler's third law of orbital motion. This same formula can be applied here:

$$r_{\text{planet}}(\text{km}) = 21.6 \times \sqrt[3]{M_{\text{star}}P^2}, \tag{28.2}$$

where M_{star} is the mass of the host star in Earth-units (see Table 28.1) and P is the exoplanet's orbital period in seconds, given in this activity's introductory paragraph.

3. (a) Compute Kepler 21b's orbital radius r_{planet}, in kilometers, from Equation (28.2) and the relevant data values. Convert your answer from kilometers into au. (1 au $= 1.5 \times 10^8$ km.) Enter your answers in the appropriate row of Table 28.2 on the worksheet. (b) Write a statement comparing Kepler 21b's orbital radius to that of the planet Mercury in our solar system: $r_{\text{Mercury}} = 0.39$ au.

28.4 Mass and Average Density of Kepler 21b

Again we follow the scheme outlined in a previous activity, where the mass of an exoplanet M_{planet} is given by the center-of-mass formula:

$$M_{\text{planet}} = \left(\frac{r_{\text{star}}}{r_{\text{planet}}}\right) \times M_{\text{star}}, \tag{28.3}$$

where M_{star} is listed in Table 28.1 and r_{planet} was derived above. The remaining unknown on the right side of the formula, r_{star}, is quantified by reintroducing from the previous activity the distance–time–velocity relation of basic physics:

$$r_{\text{star}}(\text{km}) = \frac{Pv_{\text{star}}}{2\pi}. \tag{28.4}$$

Here, the orbital period P is expressed in seconds and the orbital velocity v_{star} in kilometers per second; the resultant orbital radius r_{star} comes out in kilometers. Note: It is important to distinguish between the *planet's* orbital velocity v_{planet} and the *host star's* orbital velocity v_{star}. Both are governed by the same formula: $v = 2\pi r/P$. But whereas the planet's and star's orbital periods are identical—they remain diametrically opposite one another as they move—their orbital radii are very different: the much heavier star orbits much closer to their common center of mass than does the planet. (In fact, the star–planet center of mass is typically within the star itself.) It follows that the star's orbital velocity will be much smaller than the planet's, for it traces out a smaller circle in the same period of time. Indeed, v_{star} might be as small as a few meters per second, as opposed to the planet's kilometers per second.

The "missing link" in our plan to compute Kepler 21b's mass—v_{star} in Equation (28.4)—is determined from ultra-precise radial-velocity measurements, as described in the previous activity. These key measurements for the Kepler 21b system have recently been carried out, combining data from the High Accuracy Radial-Velocity Planet Searcher (HARPS) at the European Southern Observatory in La Silla, Chile, and the Keck Observatory's High Resolution Echelle Spectrometer (HIRES) in Hawaii. The result: $v_{star} = 2.0$ m s^{-1}. Researchers admit that this velocity could be off by as much as 0.6 m s^{-1}; it is the number that remains after the effects of the star's space motion, rotation, and cauldron-like bubbling of surface gases have been subtracted off.

4. (a) Use Equation (28.4) to obtain the radius of the star's orbit around the center of mass, r_{star}, in kilometers. (b) Use Equation (28.3) to compute the mass M_{planet} of Kepler 21b, in Earth-masses. Enter your answers in the appropriate row of Table 28.2 on the worksheet.

Now that we have both the mass and the radius of Kepler 21b, we can easily deduce its average density ρ:

$$\rho = \frac{\text{mass}}{\text{volume}} = \frac{M_{planet}}{\frac{4}{3}\pi R_{planet}^3}, \tag{28.5}$$

where the expression in the denominator is the volume of a spherical planet. To simplify the calculation, let's express all of the variables in Equation (28.5) in Earth-units. Then Kepler 21b's average density, compared to the Earth's average density ρ_{Earth} of 5.5 grams per cubic centimeter (g cm^{-3}), is given by:

$$\rho_{planet} = \frac{M_{planet(E)}}{R_{planet(E)}^3} \times \rho_{Earth}, \tag{28.6}$$

where $M_{planet(E)}$ and $R_{planet(E)}$ are expressed in Earth-masses and Earth-radii, respectively. For example, if the planet's mass is twice the Earth's mass while its radius is the same as the Earth's, then its density will be twice the Earth's density: 11 g cm^{-3}. Or if the planet's radius is twice the Earth's radius while its mass is the same as the Earth's, then its density will be $1/(2)^3$, or 1/8, of the Earth's density: 0.7 g cm^{-3}.

5. Use Equation (28.6), along with the above examples and the relevant data in Table 28.2, to derive Kepler 21b's average density, in g cm^{-3}. Enter your answer in the appropriate row of Table 28.2 on the worksheet.

6. Kepler 21b is an example of what exoplanet researchers term a *super-Earth*. Justify this characterization using the radius, mass, and density results you derived for this exoplanet.

28.5 Surface Temperature of Kepler 21b

In a previous activity, we determined the surface temperature of an airless Earth, based on an equilibrium between the incoming solar power and the outgoing energy reflected from the Earth's surface. We applied this principle to other planets, using the inverse square law to recalibrate the solar illumination for the distance of each planet. Now we do the same for the exoplanet Kepler 21b to gain a sense of what it might be like to stand on this alien world. Our expectation, given Kepler 21b's proximity to its central star, is one of searing heat and parched landscape.

The key equation from that previous activity quantifies the balance of a planet's incoming and outgoing energy:

$$T = 45.82 \times \sqrt[4]{(1 - a) \times S_{star}}, \tag{28.7}$$

where T is the surface temperature on the Kelvin scale (do not confuse with the transit duration T above); a is the planet's albedo, or reflectivity; and S_{star} is the "stellar constant," the solar-constant analog for Kepler 21b. This stellar constant is defined in the same way as the solar constant:

$$S_{star} = \frac{L_{star}}{4\pi r_{planet}^2}. \tag{28.8}$$

The formula expresses the fact that the star's luminous energy spreads out over a spherical area as it emanates through space. S_{star} represents the radiant power per square meter for a sphere the size of the planet's orbit. We compute S_{star} through comparison to the solar constant, 1370 W m^{-2} at the Earth. For example, if the star's luminosity L_{star} is twice that of the Sun, S_{star} at a distance of 1 au would be twice 1370 W m^{-2}. Or if the planet's orbital radius r_{planet} is half that of Earth (1/2 au), then S_{star} would be $1/(1/2)^2$, or 4 times 1370 W m^{-2}.

7. Considering the above examples, use Equation (28.8) and the relevant data in Tables 28.1 and 28.2 to calculate the stellar constant S_{star} at the exoplanet Kepler 21b.
8. (a) Apply Equation (28.7) to compute the surface temperature T of Kepler 21b on the Kelvin scale. Assume that the planet has an Earth-like albedo, that is, $a = 0.30$. Be sure to take the fourth root, not the square root, of the expression in Equation (28.7). (b) Convert your answer to the previous part into centigrade degrees. (°C = K − 273.) Enter your results into the appropriate row of Table 28.2 on the worksheet.

In light of Kepler 21b's physical characteristics, summarized in Table 28.2, researchers have determined that this weighty planet is too close to its host star to sustain any atmosphere. It is a dense, rocky world, unlike the gas-giant Jupiter, yet also unlike our familiar Earth. So hot is Kepler 21b's environment that the planet's outermost hundred kilometers are probably molten. This global sea of lava would be bombarded by a ceaseless blast of ultraviolet light and X-rays from the star. Whether the planet formed in place or somehow migrated inward to this hellish zone is unknown. Whichever, Kepler 21b's fiery example stands in stark contrast to the life-nurturing paradise that is our Earth.

Worksheet, Activity 28: The Search for Exoplanets: Transit Method (II)

Name _____

1. $f =$ _____

2. (a) Enter the answer in Table 28.2 below.

   (b) _____

3. (a) Compute the answer below and enter in Table 28.2.

   (b) _____

4. (a) $r_{star} =$ _____ km

   (b) Compute the answer below and enter in Table 28.2.

5. Compute the answer below and enter in Table 28.2.

6. _____

   _____

7. $S_{star} = $ _____ W m^{-2}

8. (a) Compute the answer below and enter in Table 28.2.

    (b) Compute the answer below and enter in Table 28.2.

Table 28.2. Data for Exoplanet Kepler 21b

Orbital period P	2.7858 days = 2.41×10^5 s
Planet radius R_{planet}	_____ Earth-radii
Orbital radius r_{planet}	_____ km = _____ au
Planet mass M_{planet}	_____ M_{Earth}
Average density ρ	_____ g cm^{-3}
Surface temperature T	_____ K = _____ °C

Introduction to Stars and Planets
An activities-based exploration
Alan Hirshfeld

Postscript

Having reached the end of these activities, we consider our lot as explorers of the universe. We cling to our home world, unable as yet to set foot beyond the solar system, our fragile bodies vulnerable to the extremes of temperature, pressure, radiation, and gravity that characterize the wider realm of stars, planets, even outer space itself. But our physical rootedness to the planet has anchored neither our imagination nor our drive to expand the horizons of knowledge. To the literal eye, you have been sitting at a desk, scraping a slender, graphite rod across a sheet of fibrous material, laying out strings of alphanumeric symbols. To the metaphorical eye, you have traveled backward and forward in time, plumbed the depths of stars, and hovered above the roiling lava-ocean of a distant exoplanet, all courtesy of the wondrous biochemical dynamo lodged inside your head. Science paves the way for such journeys, opening new vistas on the marvelous workings of nature.

Introduction to Stars and Planets
An activities-based exploration
Alan Hirshfeld

Appendix A

A.1 A Taste of Trigonometry

There are many applications in astronomy that involve right triangles, one of whose angles measures 90°. Trigonometry describes the mathematical relationship between the measures of the angles and the lengths of the sides of a right triangle. It does this by introducing a number of trigonometric functions.

The basic trigonometric functions of a right triangle are *sine*, *cosine*, and *tangent*, abbreviated *sin*, *cos*, and *tan* (which rhyme with the words *line*, *dose*, and *can*, respectively). For example, the sine of a 30° angle is written sin (30); the cosine of the angle θ (Greek letter *theta*) is written cos (θ). Each trig function is defined in terms of the lengths of two of a triangle's three sides: the side opposite the angle; the side adjacent to the angle; and the hypotenuse, which is the longest side. Here are the definitions, as illustrated in Figure A1:

   sin θ = opposite side/hypotenuse = a/c,
   cos θ = adjacent side/hypotenuse = b/c,
   tan θ = opposite side/adjacent side = a/b.

For example, if side a is 3 inches long, side b is 4 inches long and the hypotenuse, side c, is 5 inches long, then sin (θ) = $^3/_5$ = 0.6; cos (θ) = $^4/_5$ = 0.8; and tan (θ) = $^3/_4$ = 0.75. For such a triangle, the expression 5 sin (θ) = (5) × (0.6) = 3; and 10 tan (θ) = (10) × (0.75) = 7.5.

Alternatively, if you are given the measure of the angle θ instead of the lengths of the sides, you can use your calculator to compute the numerical value of a trig function. The keys on your calculator to do this are most likely labeled **SIN**, **COS**, and **TAN**.

Sometimes you are given, say, the sine of an angle and you have to figure out the value of the angle itself. For example, what is the angle θ whose sine is 0.6? In a sense, you are trying to "undo" the trig function to obtain the angle. For this purpose, your calculator has keys commonly labeled **SIN⁻¹**, **COS⁻¹**, and **TAN⁻¹**,

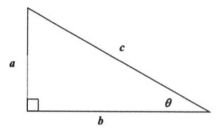

Figure A1. Right triangle with sides *a* and *b*, hypotenuse *c*, and angle *θ* labeled.

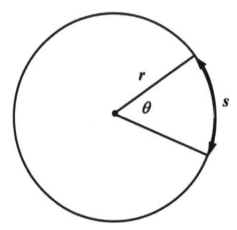

Figure A2. Sector of a circle, showing the radius *r*, arc length *s*, and enclosed angle *θ*.

which are read *inverse sine, inverse cosine,* and *inverse tangent.* You might have to first press a key labeled **2nd** (for *2nd-function*) to activate these particular keys. For practice, use your calculator to find the angle whose (a) tangent equals 1; (b) whose cosine equals 0.3.

A final word about trig functions: for the activities in this book, your calculator must be set up to accept and display angles in units of degrees, not in units called radians. On some calculators, there is a key reserved for this purpose; on other calculators, the selection is made by pressing the MENU or MODE key. If you need help, instructions for most common calculators exist online.

A.2 Arcs and Angles

In astronomy, there is frequent call for utilizing the geometry of an arc or, equivalently, a sector of a circle (basically, a "piece of pie"). What is the mathematical relationship between the angle enclosed by an arc—its *angular width*—and the length of the arc itself? In Figure A2, *s* is the length of the arc, *r* is the radius of the arc, and *θ* is the angular width of the arc.

The quantities depicted in Figure A2 are related by the geometrical formula:

$$s = (r \times \theta)/57.3, \tag{A.1}$$

where r and s are expressed in units of length (meters, inches, light-years, etc.) and θ is expressed in units of degrees. In astronomical applications, we might use Equation (A.1) to deduce, say, the true diameter s of a celestial object if its angular width θ and distance r are known. (If r is large and θ small, as they tend to be in astronomy, then the slight curvature of the arc s is indistinguishable from a celestial object's linear diameter.)

Equation (A.1) can be rewritten as $r = (57.3 \times s)/\theta$, with which we can compute the distance r of a celestial object if its true diameter s and angular width θ are known. Or if the equation is recast as $\theta = (57.3 \times s)/r$, we can deduce an object's angular span θ in the sky if its true diameter s and distance r are known. For example, a 4 cm-wide ping pong ball (s) spans an angle of about $3°$ (θ) if viewed at arm's length r, about 75 cm.

A.3 Powers-of-ten Notation

Given the enormous scale of sizes, masses, distances, densities, and timespans, as well as the seeming infinitude of objects in the universe, astronomers deal routinely with very large numbers. Rather than writing out long strings of digits, scientists have developed a shorthand method to express large numbers: *powers-of-ten notation*, also known as *scientific notation*. In this system, every number consists of a two-part mathematical expression: a coefficient between 1 and 10; and a multiplier that is a power of 10, that is, 10 raised to an integer exponent. The powers of ten are:

$10 = 10^1$
$100 = 10^2$
$1000 = 10^3$
$10,000 = 10^4$
$100,000 = 10^5$
$1,000,000 = 10^6$, and so on.

To convert the number 510,000 to powers-of-ten notation, move the decimal point from its current position—it's implied after the rightmost zero—to a position between the first and second digits of the number, in this case 5.10000. Count the number of decimal places after (to the right of) the decimal point, and that becomes the exponent, or power of ten. Thus, 510,000 can be written 5.1×10^5. Similarly, the radius of the Earth's orbit, 150,000,000 km, is written 1.5×10^8 km. And the Earth's mass, about 5,973,600,000,000,000,000,000,000 kg, abbreviates to 5.9736×10^{24} kg.

By the way, on many calculator displays, the exponent portion of a powers-of-ten number is indicated by the letter E. Thus, 6.2×10^{20} might appear on your calculator as 6.2 E20. To input a powers-of-ten number into your calculator, key in the coefficient portion of the number, then press the exponent key (labeled **EE**, **EXP**, or perhaps ^), followed by the exponent.

CPSIA information can be obtained
at www.ICGtesting.com
Printed in the USA
BVHW060323081121
621028BV00006B/161